D0024702

Bioprocessing Pipelines:
Rheology and Analysis

Bioprocessing Pipelines:
Rheology and Analysis

James F. Steffe, Ph.D., P.E.

and

Christopher R. Daubert, Ph.D.

Freeman Press
2807 Still Valley Drive
East Lansing, MI 48823-2351
USA

James F. Steffe, Ph.D., P.E.
Dept. of Biosystems and Agricultural Engineering
Dept. of Food Science and Human Nutrition
Michigan State University
East Lansing, MI 48824-1323

steffe@msu.edu

and

Christopher R. Daubert, Ph.D.
Dept. of Food Science
North Carolina State University
Raleigh, NC 27695-7624

cdaubert@ncsu.edu

Copyright © 2006 (All rights reserved) by James F. Steffe and Christopher R. Daubert. No part of this work may be reproduced, stored in a retrieval system, or transmitted, in any form or by any means, electronic, mechanical, photocopying, recording, or otherwise, without the written permission of the authors.

Printed on acid free paper in the USA.

Library of Congress Control Number: 2005905876

ISBN-10: 0-9632036-2-2
ISBN-13: 978-0-9632036-2-5

Freeman Press
2807 Still Valley Drive
East Lansing, MI 48823
USA

Preface

This book grew from the professional experience of the authors and recommendations from industrial colleagues. Our examination of the trade literature revealed scant information on non-Newtonian fluids, as well as inaccurate descriptions of rheological behavior pertaining to pipeline design calculations. In addition, there is a great deal of published research (some with our names on it) dealing with the rheology of biological fluids and the analysis of pipeline processes; however, this information has not been distilled and synthesized into a form that is useful for attacking practical bioprocessing problems. We hope our work shines new light on the area, and provides valuable tools for every day practice.

The intended audience for this book is students, technologists, and practicing engineers interested in processing biological fluids, primarily food and pharmaceutical fluids. These materials are subjected to a variety of mechanical forces and thermal treatments during processing. Our work is designed for self-study; and, after sufficient effort, we hope readers will be able to: 1) Explain the basic principles of fluid rheology needed to examine bioprocessing pipeline design problems; 2) Determine (using an appropriate instrument) the rheological properties of biological fluids needed to calculate pipeline design parameters; 3) Solve pumping problems (for Newtonian and non-Newtonian fluids) using the mechanical energy balance equation as the framework for the analysis; 4) Characterize the shear and thermal treatments given to biological materials in fluid processing systems.

In the 14th century, William of Occam said *"Pluralitas non est ponenda sine necessita"* which, in modern terms could be interpreted as "keep things simple." We have embraced this principle. Rheology is a complex topic, but we have simplified it using Occam's razor to cut away assumptions, theories and models that are not needed to characterize fluids for the purpose of pipeline analysis. Also, we have eliminated derivations of equations to just present the useful results. Hopefully, this approach will quickly allow our readers to find meaningful solutions to practical pipeline design problems. More detailed information on rheological techniques and data interpretation may be found in ***Rheological Methods in Food Process Engineering, Second Edition*** (1996, Freeman Press), by J.F. Steffe. This work is available at no charge: www.egr.msu.edu/~steffe/freebook/offer.html

We – in our multiple roles as authors, engineers, and professors – are committed to protecting the environment and to the responsible use of natural resources. Since our work is produced on paper, we are concerned about the future of the world's remaining endangered forests and the environmental impacts of paper production; and we are committed to furthering policies that will support the preservation of endangered forests globally and advance best practices within the book and paper industries. Furthermore, we encourage publishers, printers, and our fellow authors to endorse the paper use recommendations of the Green Press Initiative (www.greenpressinitiative.org). The paper used in this book meets those recommendations.

J. F. Steffe
C. R. Daubert

Table of Contents

Preface v
Nomenclature ix

1 Rheological Properties of Biological Fluids 1
1.1 Viscosity and Empirical Fluid Parameters 1
1.2 Useful Fluid Models 2
1.3 Time-Independent versus Time-Dependent Fluids 6
1.4 Shear Stress and Shear Rate in a Pipeline 8
1.5 Shear-Thinning Fluid Behavior in a Pipeline 9
1.6 Shear Rate Selection for Rheological Data Collection 10
1.7 Influence of Temperature on Rheological Behavior 12

2 Pipeline Rheology 15
2.1 Characterizing Fluids for Pipeline Design 15
2.2 Concentric Cylinder Viscometry 15
2.3 Mixer Viscometry 19
2.3.1 Characterizing the Mixer Viscometer 20
2.3.2 Finding Properties of Power Law Fluids 27
2.3.3 Brookfield Viscometers with Disk Spindles 30
2.4 Cone and Plate, and Parallel Plate Viscometers 33

3 Energy and Head Loss Calculations 35
3.1 Mechanical Energy Balance Equation 35
3.2 System Curves (Process Requirements) 38
3.3 Finding Work and Head 39
3.4 Pump Curves (Centrifugal Pumps) 45
3.5 Net Positive Suction Head (Available and Required) 46

4 Fanning Friction Factors 49
4.1 Friction Factors 49
4.2 Newtonian Fluids 49
4.3 Power Law Fluids 51
4.4 Tube Flow Velocity Profiles 55

5 Friction Loss Coefficients 61
5.1 Losses in Standard Valves and Fittings 61
5.2 Losses in Equipment Based on Data for Water 63

6 Handling Shear-Sensitive Fluids 72
6.1 Shear-Sensitive Fluids 72
6.2 Shear Work (W_s) 73
6.3 Shear Power Intensity (S) 75
6.4 Critical Values of W_s and S 77
6.5 Shear-Sensitive Particulates 81
6.6 Scale-Up Calculations 82

7 Thermal Processing of Biological Fluids 85
7.1 Death Kinetics of Microorganisms 85
7.2 The General Method 89

8 Example Problems 96
8.1 Comparison of Newtonian and Shear-Thinning Fluids 96
8.2 Herschel Bulkley and Casson Equations to Power Law 99
8.3 Concentric Cylinder Data for Ice Cream Mix 102
8.4 Determination of the Mixer Coefficient 106
8.5 Mixer Viscometry Data for Pasta Sauce 107
8.6 Calculating Pressure Drop with Effective Viscosity 110
8.7 Generating a System Curve for Pumping Cream 113
8.8 Positive Displacement Pump for Pulpy Fruit Juice 120
8.9 Pumping a Shear Sensitive Fluid (Cream) 127
8.10 Shear Power Intensity in a Centrifugal Pump 132
8.11 Lethality of Pasteurization Process 133

9 Appendices 136
9.1 Conversion Factors and Greek Alphabet 136
9.2 Rheological Properties of Biological Fluids 138
9.3 Stainless Steel Tubing and Pipe Diameters 140
9.4 Stainless Steel (304 and 316) 141
9.5 Properties of Saturated Water 143
9.6 Enthalpy of Saturated Steam 145
9.7 Viscosity (mPa s or cP) of Water 147
9.8 Gallons of Water per 100 feet of Tubing 148
9.9 Affinity Laws for Centrifugal Pumps 148
9.10 Equations for Bingham Plastic Fluids in Tube Flow 149
9.11 Fanning Friction Factors for Power Law Fluids 152
9.12 Friction Loss Coefficients: 3-k equation 156

Index 158

Nomenclature

This book uses a combination of English and SI units. Although the authors prefer the SI system, the mixed units presented in the text represent the current practice of the US food and pharmaceutical industries.

A	constant, Pa s
A	area, m^2
A	constant, dimensionless
B	constant, dimensionless
C	constant, dimensionless
d	impeller diameter, m
D	decimal reduction time, min
D	diameter, m
D_{inch}	diameter, inch
D_o	decimal reduction time at 250°F, min
D_h	hydraulic diameter, m
E_a	activation energy for flow, cal / (g-mole)
f	Fanning friction factor, dimensionless
F	thermal death time, min or s
F	viscous energy loss (friction loss) per unit mass, J kg^{-1}
F	force, N
F_o	thermal death time at 250°F, min
g	gravitational acceleration, 9.81 m s^{-2}
h	height of bob or separation between parallel plates, m
h_o	height added to bob for end correction, m
h'	height defined by Fig. 2.1, m
h''	height defined by Fig. 2.1, m

H_p	total pump head, m
H_s	total system head, m
k	rate constant for microbial inactivation, min^{-1}
k'	mixer viscometer constant, rad^{-1}
k''	mixer coefficient, rad m^{-3}
k_1	constant, dimensionless
k_f	friction loss coefficient, dimensionless
$k_{\dot{\gamma}}$	shear rate conversion factor, min rev^{-1} s^{-1}
k_σ	shear stress conversion factor, Pa
k_∞	constant, dimensionless
K	consistency coefficient, Pa s^n
L	length, m
L_e	equivalent length, m
LR	lethal rate, dimensionless
M	torque, N m
m	mass, kg
$\dot{m}$	mass flow rate, kg s^{-1}
n	flow behavior index, dimensionless
N	angular velocity, rev min^{-1}
N	number of microorganisms
N_{He}	Hedstrom number, dimensionless
N_0	initial number of microorganisms
N_{Po}	power number, dimensionless
N_{Re}	Reynolds number (for Newtonian fluids), dimensionless
$N_{\text{Re,B}}$	Bingham Reynolds number, dimensionless
$N_{\text{Re},I}$	impeller Reynolds number, dimensionless
$N_{\text{Re},PL}$	power law Reynolds number, dimensionless

$N_{\mathrm{Re},\,PL,I}$	power law impeller Reynolds number, dimensionless
$(NPSH)_A$	net positive suction head available, m
$(NPSH)_R$	net positive suction head required, m
P	pressure, Pa
P_v	vapor pressure, Pa
r	radius (variable) or radius of curvature, m
R	radius, m
R'	universal gas constant, 1.987 cal / (g-mole K)
R_b	bob radius, m
R_c	cup radius, m
R_s	shaft radius, m
R_o	critical radius, m
RKT	relative kill time, dimensionless
S	shear power intensity, J s^{-1} m^{-3}
SV	sterilizing value, dimensionless
t	time, min or s
T	temperature, °C, °F or K
T_r	reference temperature, °C, °F or K
Q	volumetric flow rate, m^3 s^{-1}
u	velocity, m s^{-1}
$\overline{u}$	volumetric average (or mean) velocity, m s^{-1}
$u_{\max}$	maximum velocity, m s^{-1}
u^+	dimensionless velocity
V	volume, m^3
W	shaft work input per unit mass, J kg^{-1}
W_s	shear work per unit mass, J kg^{-1}
z	elevation, m

z	temperature change for a one log reduction in D, °F or °C
α	R_c/R_b, dimensionless
α	kinetic energy correction factor, dimensionless
α	constant, dimensionless
$\dot{\gamma}$	shear rate, s^{-1}
$\dot{\gamma}_a$	average shear rate, s^{-1}
$\dot{\gamma}_b$	shear rate at the bob, s^{-1}
$\dot{\gamma}_{\max}$	maximum shear rate, s^{-1}
$\dot{\gamma}_{\min}$	minimum shear rate, s^{-1}
$\dot{\gamma}_R$	shear rate at rim of a parallel plate sensor, s^{-1}
Δ	constant, dimensionless
ΔP	pressure loss or change in pressure, Pa
ε	surface roughness, m
η	apparent viscosity, Pa s
η	pump efficiency, dimensionless
η_r	reference apparent viscosity, Pa s
θ	angle, degrees
μ	absolute viscosity or viscosity of a Newtonian fluid, Pa s
μ_{pl}	plastic viscosity, Pa s
μ_r	reference Newtonian viscosity, Pa s
ν	kinematic viscosity, cSt or $mm^2\ s^{-1}$
ρ	density, $g\ cm^{-3}$ or $kg\ m^{-3}$
σ	shear stress, Pa
σ_a	average shear stress or representative average, Pa

σ_b	shear stress at the bob, Pa
σ_o	yield stress, Pa
σ_{max}	maximum shear stress, Pa
σ_R	shear stress at rim of a parallel plate sensor, Pa
Φ	power, W or J s^{-1}
Ω	angular velocity, rad s^{-1}

1 Rheological Properties of Biological Fluids

1.1 Viscosity and Empirical Fluid Parameters

Rheology is the science of the deformation and flow of matter. Rheological properties of biological fluids can vary greatly, even within the same general product categories such as applesauce, ketchup or chocolate; hence, it is important that rheological behavior be carefully evaluated for all new products. Flow behavior can be broadly characterized in terms of two measurements: measurement of absolute properties (needed for pipeline design) that are independent of the measuring instrument, and empirical measurements (often useful in quality control applications) where results depend completely on the physical characteristics of the measuring device. Without good rheological data there is little accuracy in pipeline design calculations.

Errors and misconceptions can be found in the rheological property data and descriptions provided by various companies. A few issues demand immediate discussion:

- The word "viscosity" is used to describe many different properties in rheology and most of them have no application in the analysis of bioprocessing pipelines. Examples (and this is not an exhaustive list) include the following: extensional viscosity, inherent viscosity, intrinsic .viscosity, reduced viscosity, specific viscosity, complex viscosity, and dynamic viscosity. Clearly, one must be careful when using the word viscosity to characterize fluid properties.

- Describing materials using the word "viscosity" implies the fluid in question is Newtonian. Kinematic viscosity is also common, and also implies Newtonian behavior.
- Non-Newtonian fluids cannot be properly characterized with a single measure of viscosity in centipoise, or any other set of units. Examples of empirical units that should not be used (but have been suggested or are currently used) to characterize non-Newtonian fluids for pipeline design calculations include the following: Saybolt Universal Seconds, Degrees Engler, Dupont Parlin, Krebs, Redwood viscosity, MacMichael, RVA (Rapid Visco Analyser) viscosity, and Brabender viscosity. There is also a long list of empirical instruments that are inappropriate for determining pipeline design characteristics of non-Newtonian fluids including dipping cups (e.g., Parlin Cup, Ford Cup, Zahn Cup), rising bubbles, falling balls, rolling balls; and some food industry specific instruments such as the Brabender Viscoamlyograph, Mixograph, Farinograph, Adams Consistometer, and the Bostwick Consistometer.

1.2 Useful Fluid Models

A fluid model is a mathematical equation that describes flow behavior; and appropriate models are determined from statistical curve fitting (typically, linear regression analysis) of experimental data. A fluid is Newtonian if the observed behavior for that substance responds independently of time, displays a linear relationship between shear stress and shear rate, and has no yield stress. All other fluid scenarios are considered non-Newtonian. This section will focus on fluids that are time-independent, meaning they have no memory. Time-dependent behavior is discussed in the next section.

The flow behavior of fluids is characterized with the experimentally determined relationship between shear stress and shear rate. These parameters can be explained by considering the steady, simple shear behavior of a fluid flowing between two parallel plates (Fig. 1.1) with a surface area (A) contacting the fluid, and plates separated by a distance, h. The lower plate is fixed (the velocity equals zero, $u = 0$), and the upper plate moves at a maximum velocity equal to u. A force (F) is required to maintain the velocity of the upper plate. Using these variables, shear stress (σ) is defined as force divided by the area:

$$\sigma = \frac{F}{A} \tag{1.1}$$

with units of pressure, typically defined as a Newton per square meter or a Pascal. Shear rate $\left(\dot{\gamma}\right)$, corresponding to the shear stress, is the velocity of the upper plate divided by the distance separating the plates:

$$\dot{\gamma} = \frac{u}{h} \tag{1.2}$$

Since velocity has units of meters per second and height has units of meters, the units of shear rate are reciprocal seconds, s^{-1}. Viscometers are instruments that collect experimental data to determine shear stresses and shear rates under varying conditions.

To describe material behavior, shear stress must be related to shear rate. Many fluids can be described as Newtonian because rheological data show the relationship between shear stress and shear rate to be linear:

$$\sigma = \mu\dot{\gamma} \tag{1.3}$$

where μ is the absolute viscosity of the fluid. Biological fluids that fall into this category include water, clear fruit juice, honey, alcoholic beverages, soft drinks, and liquid oils (olive oil, corn oil, etc.). Typical units of absolute viscosity are Pascal second or centipoise (1 Pa s = 1000 cP). Viscosity conversion factors are provided in Appendix 9.1.

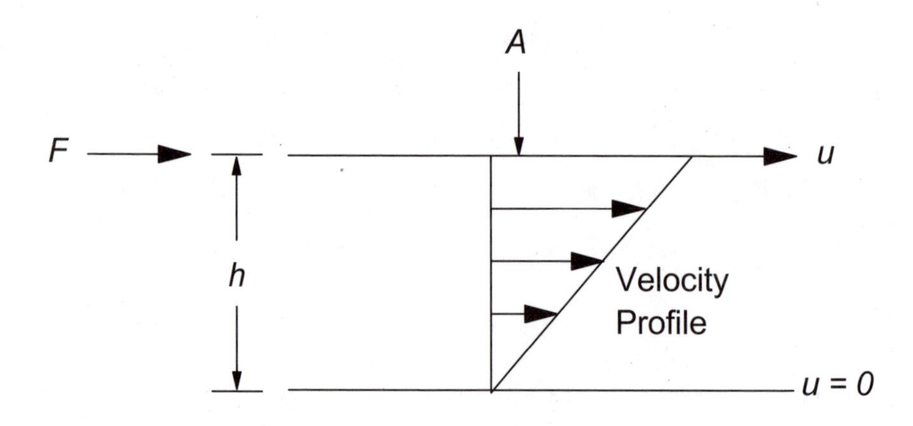

Figure 1.1. Fluid flow between parallel plates.

The flow behavior of Newtonian fluids is sometimes described in terms of the kinematic viscosity, defined as the absolute viscosity divided by the density:

$$\nu = \frac{\mu}{\rho} \tag{1.4}$$

The usual unit for kinematic viscosity is centistokes. Using water at 20°C, for example,

$$\nu = \frac{\mu}{\rho} = \left[\frac{1\,cP}{1\,g\,cm^{-3}} \right] = \left[1\,cSt \right] \tag{1.5}$$

Fluids that do not obey the Newtonian model given by Eq. (1.3) are, by definition, non-Newtonian. For the purpose of pipeline analysis, the majority of non-Newtonian fluids can be successfully described with the power law model:

$$\sigma = K\dot{\gamma}^n \tag{1.6}$$

where K is the consistency coefficient (units of Pa s^n) and n is the dimensionless flow behavior index. The Newtonian model is a special case of the power law model: when n = 1.0, the equation collapses into the Newtonian model and $K = \mu$. Values of $0 < n < 1$ indicate shear-thinning behavior (very common), and $n > 1$ indicate shear-thickening behavior (uncommon). Synonyms for shear-thinning and shear-thickening are pseudoplastic and dilatent, respectively. Shear-thinning and shear-thickening are the preferred terms since they are more descriptive of the fluid behavior being characterized. Also note, there is nothing "pseudo" about shear-thinning fluids, and shear-thickening fluids do not "dilate." Examples of shear-thinning fluids include concentrated or pulpy fruit juices, fruit and vegetable purees, puddings, and weak gel solutions. Typical values of the consistency coefficient and the flow behavior index for biological fluids are given in Appendix 9.2.

The power law model may be described in terms of the apparent viscosity (η) defined as shear stress divided by shear rate:

$$\eta = \frac{\sigma}{\dot{\gamma}} = \frac{K\dot{\gamma}^n}{\dot{\gamma}} = K\dot{\gamma}^{n-1} \tag{1.7}$$

Apparent viscosity varies with shear rate and depends on the numerical values of K and n. If a fluid is Newtonian, the apparent viscosity and the Newtonian viscosity are equal. Eq. (1.6) and Eq. (1.7) are applied to blueberry pie filling and honey in Example Problem 8.1. Apparent viscosity should not be confused with effective viscosity -- the two are compared in Example Problem 8.6.

The vast majority of fluid pumping problems can be solved using the Newtonian or power law models. This is beneficial because the equations required for pipeline analysis of these materials are well developed. The power law model is very appropriate for most non-Newtonian fluids and generally includes materials that may be described as pourable or easily "spoonable." Also, the power law equation is often a good approximation for the behavior of fluids that have a significant yield stress (the minimum stress required to initiate flow) such as tomato paste. Very thick pastes, having very large yield stresses, may require more complex rheological models, such as the Bingham plastic (see Appendix 9.10) equation. Sometimes, however, more complex equations are applied to fluids to increase curve fitting accuracy even though the power law model is acceptable for the purpose of pipeline design. Example Problem 8.2 illustrates a case where Herschel-Bulkley and Casson equations are given for baby food, then converted to the power law model to provide more useful pipeline design properties.

1.3 Time-Independent versus Time-Dependent Fluids

Newtonian and power law fluids are considered time-independent because they have no memory. The rheological properties of these materials are unaffected by mechanical effects introduced from operations such as mixing or pumping. Measuring the viscosity of water or olive oil, for example, will be the same no matter how long they are mixed or left undisturbed. On the other hand, a weak gel (pudding or lotion) may display a diminished consistency with mixing due to the destruction of weak structures within the material.

Time-dependent fluids fall into two categories: thixotropic fluids that thin with time, and rheopectic (also called anti-thixotropic) fluids that thicken with time. These terms can be explained using ketchup as an example fluid. Tomatoes are subjected to various operations (washing, peeling, crushing, and pulping) before the pureed product is placed in a container for commercial distribution. When pumping the fluid at the point of manufacture, the product is completely broken down, and the ketchup behaves as a shear-thinning (time-independent) fluid. If the bottled ketchup remains undisturbed for a period of weeks, the fluid thickens by forming a weak gel structure (caused mainly by the presence of pectin in tomatoes) within the product. Subjecting the aged ketchup to shearing -- due to mechanical agitation induced by stirring or shaking the container -- breaks down the weak structure and makes the product more pourable. The time and extent of mechanical energy input determines the degree of time-dependent thinning. Providing the ketchup with a sufficient level of mechanical degradation causes the fluid to reestablish a time-independent behavior similar to what it had at the point of manufacture. If the degradation process was controlled within the confines of a viscometer, the thixotropic behavior of the ketchup could be qualitatively evaluated. A rheologist might make these measurements to study the shear-sensitivity, shelf-stability, or consumer acceptability of the ketchup. When selecting pumps for a processing line, ketchup can be treated as a time-independent fluid and only the power law fluid behavior need be considered.

1.4 Shear Stress and Shear Rate in a Pipeline

Experimental data of shear stress and shear rate are required to determine the rheological properties of a fluid. An appropriate range of these parameters must be selected for data to be meaningful. When a power law fluid moves through a pipeline under laminar flow conditions (criterion to establish laminar flow is discussed in Section 4.3), the following equation describes the relationship between the pressure drop inducing flow (ΔP) and the resulting volumetric flow rate (Q):

$$\frac{\Delta P\,R}{2L} = K\left[\left(\frac{3n+1}{4n}\right)\left(\frac{4Q}{\pi R^3}\right)\right]^n \tag{1.8}$$

Comparing Eq. (1.8) to Eq. (1.6), one can write expressions for the maximum shear stress and the maximum shear rate found in the pipe:

$$\sigma_{\max} = \frac{\Delta P\,R}{2L} \tag{1.9}$$

$$\dot{\gamma}_{\max} = \left(\frac{3n+1}{4n}\right)\left(\frac{4Q}{\pi R^3}\right) \tag{1.10}$$

Eq. (1.10) is used in Example Problem 8.1 to calculate the maximum shear rate found in pumping blueberry pie filling.

The maximum values given above occur at the inside wall of the pipe where the value of the radius variable (r) is equal to R, the inside radius of the pipe. The minimum value of the shear stress occurs at the center of the pipe ($r = 0$), and varies linearly with the radius:

$$\sigma = \left(\frac{\Delta P\,r}{2L}\right) \tag{1.11}$$

The minimum value of the shear rate in a pipe is zero and it also occurs at the center of the pipe ($r = 0$), but varies non-linearly (unless the fluid is Newtonian so $n = 1$) with the radius:

$$\dot{\gamma} = \left[\left(\frac{3n+1}{4n}\right)\left(\frac{4Q}{\pi R^3}\right)\right]\left(\frac{r}{R}\right)^{\frac{1}{n}} \tag{1.12}$$

The above equations are only valid for laminar flow because shear rate in a pipeline is not easily defined for turbulent flows.

1.5 Shear-Thinning Fluid Behavior in a Pipeline

To better understand the physical meaning of shear-thinning ($0 < n < 1$), consider the relationship between pressure drop and volumetric flow rate during pumping. The pressure drop in a pipe for the laminar flow of a Newtonian fluid is described by the Poiseuille-Hagen equation:

$$\Delta P = \left(\frac{8L\mu}{\pi R^4}\right)Q \tag{1.13}$$

Similarly, the pressure drop in a pipe for the laminar flow of a power law fluid is

$$\Delta P = \frac{2KL}{R}\left[\left(\frac{3n+1}{4n}\right)\left(\frac{4}{\pi R^3}\right)\right]^n Q^n \tag{1.14}$$

In the case of a Newtonian fluid, Eq. (1.13) shows the pressure drop is directly proportional to the flow rate; doubling the flow rate (Q) results in a doubling of the pressure drop over a length of pipe equal to L. In the case of a shear-thinning fluid (Eq. (1.14)), the pressure drop is proportional to the flow rate raised to the power n; doubling the flow rate, causes the pressure drop to increase by a factor of 2^n. If $n = 0.5$, for example, the pressure drop would increase by a factor of 1.41.

Now, reconsider the concept of shear-thinning. Doubling Q with the Newtonian fluid doubled ΔP; however, doubling Q with the shear-thinning fluid having a flow behavior index of 0.5 only increased the pressure drop to $1.41\Delta P$. Since doubling Q did not double ΔP, it appears that increasing the shear rate (a consequence of increasing Q) caused the non-Newtonian fluid to appear less viscous, i.e., thinner. This phenomenon is the result of power law fluid behavior: the rheological properties of the material (K and n) have not changed.

1.6 Shear Rate Selection for Rheological Data Collection

Experimental data collected to determine rheological properties for the purpose of pipeline design should cover approximately the same shear rate range found in the pipe. Theoretically, it is possible to match shear stresses instead of shear rates, but this is usually impracticable due to our limited knowledge of the pressure drop. In practice, estimating the shear rate range for the laminar or turbulent flow regimes is based on pipeline equations for laminar flow. One should also be aware that actual shear rates may vary a great deal in different pieces of equipment that constitute the total pipeline system. The maximum shear rates found in strainers and partially open pneumatic valves, for example, may be much higher than the maximum shear rate found in a pipe. Also, at a constant flow rate, pipes of different diameters will have different maximum shear rates.

An upper shear rate limit can be estimated from Eq. (1.10) given values of the maximum volumetric flow rate, the radius of the pipe, and the flow behavior index. For a Newtonian fluid, the maximum shear rate is calculated as $4Q/(\pi R^3)$. To calculate the maximum shear rate for non-Newtonian fluids, this value must be multiplied by the shear rate correction factor:

$$\text{shear rate correction factor} = \left(\frac{3n+1}{4n}\right) \tag{1.15}$$

It is important to get a good estimate of the flow behavior index because the numerical value of the correction factor is strongly influenced by this rheological property. An n value of 0.2, for example, leads to a shear rate correction factor of 2 which doubles the maximum shear rate calculated using the simple Newtonian approximation. If a value of the flow behavior index is unknown – which may be the case since a shear rate range is needed for the purpose of evaluating rheological properties – one may be able to obtain a good estimate of n based on the properties of a similar product (see Appendix 9.2). The best procedure, however, is to collect preliminary rheological data to determine n before the final shear rate range is established.

Although the minimum shear rate in a pipe is zero, it is unnecessary to collect data at very low shear rates except in cases involving extremely low volumetric flow. Rheological data taken at shear rates less than 1 s^{-1} are generally not required in pipeline design work. A reasonable lower limit of the shear rate is approximately 1/10 of the maximum shear rate limit predicted with Eq. (1.10). This idea is applied in Example Problem 8.1. Maximum shear rates for some commercial and pilot plant dairy processes are given in Table 1.1.

Table 1.1. Maximum shear rates in typical pumping systems.

Product (process)	n -	flow rate (lbm/hr)	flow rate (gal/min)	D^+ (in)	$\dot{\gamma}_{max}$ (1/s)
Ice Cream Mix (pilot plant)	0.7	907	8	2	53
Milk (pilot plant)	1.0	1481	13	2	86
Ice Cream Mix (commercial)	0.7	22665	200	4	154
Process Cheese (commercial)	0.3	27,000	50	3	136
Milk (commercial)	1.0	130,000	252	4	175

D^+ = nominal tube diameter; see Appendix 9.3 for actual diameters.

1.7 Influence of Temperature on Rheological Behavior

Rheological behavior is strongly influenced by temperature, and the change in Newtonian viscosity (μ) can be modeled with the Arrhenius equation:

$$\mu = A \exp\left(\frac{E_a}{R'T}\right) \tag{1.16}$$

where E_a is the activation energy for flow, R' is the universal gas constant, and T is the absolute temperature. Eq. (1.16) can also be expressed in terms of a reference viscosity (μ_r) given at a specific reference temperature (T_r):

$$\ln\left(\frac{\mu}{\mu_r}\right)=\left(\frac{E_a}{R'}\right)\left(\frac{1}{T}-\frac{1}{T_r}\right) \tag{1.17}$$

When evaluating power law fluids, the apparent viscosity at a constant shear rate can also be modeled using the Arrhenius equation:

$$\ln\left(\frac{\eta}{\eta_r}\right)=\left(\frac{E_a}{R'}\right)\left(\frac{1}{T}-\frac{1}{T_r}\right) \tag{1.18}$$

This relationship is useful since the flow behavior index (n) is generally not a strong function of temperature. If n is not temperature dependent, the consistency coefficient (K) can also be modeled with the Arrhenius equation. Note that temperature used in the Arrhenius equation is given in degrees Kelvin, so the ratio E_a/R' must have units of degrees Kelvin.

Using the above equations to summarize the influence of temperature on flow behavior can be valuable when a range of temperatures must be considered. For example, data for honey show Newtonian behavior with viscosities ranging from 76.9 Pa s at 6.5°C, to 0.50 Pa s at 48.0°C. Collecting data at values between these numbers and fitting the data to Eq. (1.16) allowed the viscosity (in units of Pa s) to be modeled as a function of temperature:

$$\mu = A\exp\left(\frac{E_a}{R'T}\right)=5.58\left(10^{-16}\right)\exp\left(\frac{10,972}{T}\right) \tag{1.19}$$

where E_a/R' = 10,972 K. Behavior can also be modeled by rearranging Eq. (1.18):

$$\mu = \mu_r \exp\left[\frac{E_a}{R'}\left(\frac{1}{T} - \frac{1}{T_r}\right)\right] \tag{1.20}$$

or, with the substitution of a reference viscosity (3.77 Pa s) at 300 K, as

$$\mu = 3.77 \exp\left[10{,}972\left(\frac{1}{T} - \frac{1}{300}\right)\right] \tag{1.21}$$

Eq. (1.19) and (1.21) allow the viscosity of honey to be calculated at any temperature between 6.5°C and 48.0°C (278.7 K to 321.2 K). Eq. (1.19) is used to determine the flow behavior of honey at 24°C in Example Problem 8.1.

In a second study, concentrated orange juice was evaluated at temperatures of –18.8°C, -5.4°C, 9.5°C, and 29.2°C (254.4 K, 267.8 K, 282.7 K, and 302.4 K). The fluid exhibited power law behavior at each temperature with little variation in the flow behavior index (average n = 0.77) but major changes in the consistency coefficient. Modeling the consistency coefficient with the Arrhenius equation gave the following result:

$$K = 4.646\left(10^{-9}\right)\exp\left(\frac{5668}{T}\right) \tag{1.22}$$

Eq. (1.22) makes the value of K readily accessible at any temperature between –18.8°C and 29.2°C (254.4 K to 302.4 K). This example is also instructive because it shows why the Arrhenius equation is expressed using absolute temperature instead of degrees Celsius: when T = 0, Eq. (1.16) is mathematically undefined.

2 Pipeline Rheology

2.1 Characterizing Fluids for Pipeline Design

Fluid flow problems involving pipeline design and pump selection fall in the steady shear testing domain of the vast field of rheology. The most common and useful steady shear instruments for determining fluid properties for pipeline design are the rotational type viscometers. In this instrument category, concentric cylinder systems are the most applicable, and the best starting point for a pipeline rheology lab; cone and plate, and parallel plate systems can also be useful. One important disadvantage of these sensors is the significant particle size limitations, particularly for cone and plate sensors. Mixer viscometers (discussed in Sec. 2.3) can be used to evaluate flow behavior when particles are large enough to impede shear flow in conventional rotational instruments.

2.2 Concentric Cylinder Viscometry

The concentric cylinder viscometer may be described as one cylinder (the bob) placed inside a second cylinder (the cup) with the test fluid located in the annular space between them. A popular geometry, based on a German standard generated by Deutsches Institut für Normung (DIN), is illustrated in Fig. 2.1. To maintain accurate calculations, the following preferred dimensions are specified for the concentric cylinder geometry in the DIN 53019 standard: $\alpha = \frac{R_c}{R_b} = 1.0847$; $\frac{h}{R_b} = 3$; $\frac{h'}{R_b} = 1$; $\frac{h''}{R_b} = 1$; $\frac{R_s}{R_b} \leq 0.3$; $\theta = 120° \pm 1°$.

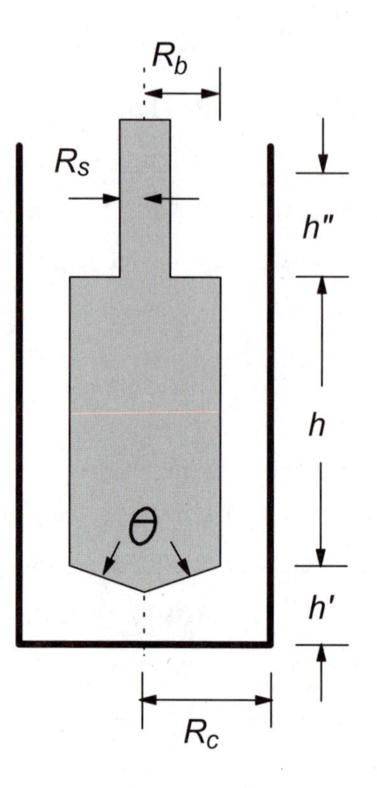

Figure 2.1. Concentric cylinder system based on the DIN 53019 standard.

where θ is the apex angle of the cone at the bottom of the inner cylinder. When testing, fluid is filled to a height of h'' (Fig. 2.1). Another popular design, based on DIN 53018, is illustrated in Fig. (2.2). This geometry includes a recessed top and bottom to minimize end-effect errors.

In the typical instrument operation of a concentric cylinder viscometer, the operator will select a rotational speed (Ω) for the bob and measure the torque (M) resisting the rotation. Data are collected at many different speeds corresponding to the shear rate range appropriate for the problem under consideration.

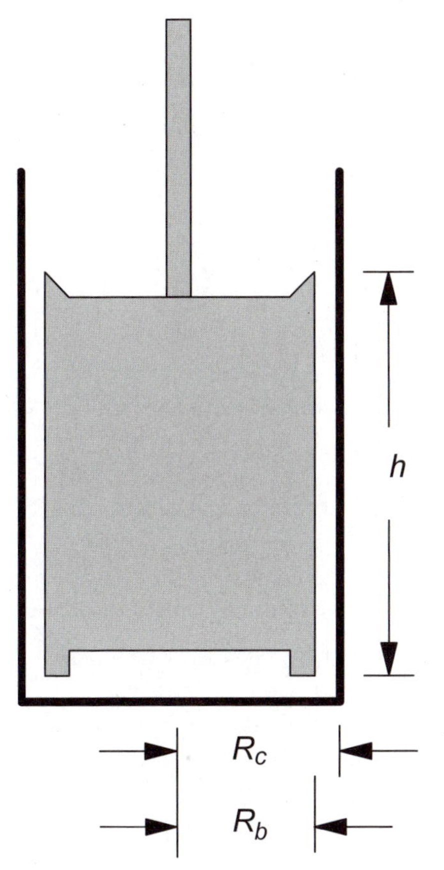

Figure 2.2. Concentric cylinder system based on the DIN 53018 standard.

Rheological properties are determined by constructing a plot of shear stress versus shear rate (called a rheogram) from the raw data. Shear stress is directly proportional to torque, and shear rate is directly proportional to rotation speed (angular velocity). The shear stress and the shear rate at the surface of the bob (σ_b and $\dot{\gamma}_b$, respectively) may be calculated as

$$\sigma_b = M\left[\frac{1}{2\pi R_b^2\left(h+h_o\right)}\right] \tag{2.1}$$

and

$$\dot{\gamma}_b = \Omega\left[\frac{2\alpha^2}{\alpha^2-1}\right] \tag{2.2}$$

where $\alpha = R_c / R_b$.

The h_o variable in the denominator of Eq. (2.1) is a fictitious addition to the height of the bob (h) accounting for the end effect on the bob: the end effect is the torque contribution caused by the top and/or bottom end faces of the bob. For the bob illustrated in Fig. 2.1 (DIN 53019), the correction is approximately 10% of the height that is parallel to the cup: $h_o = 0.1h$. Using the design illustrated in Fig. 2.2, and keeping the distance between the bottom of the cup and the bottom of the bob to a distance of approximately three times the gap width, the end effect correction is very small. With this restriction, the end effect correction is approximately 0.5 percent of the bob height: $h_o = 0.005h$.

Average values of shear stress and shear rate within the annulus may also be used. The DIN 53019 standard specifies that an average shear stress (called the representative shear stress) be determined as

$$\sigma_a = M\left[\frac{1+\alpha^2}{4\pi R_c^2\left(h+h_o\right)}\right] \tag{2.3}$$

The corresponding average shear rate (called the representative shear rate) is

$$\dot{\gamma}_a = \Omega\left[\frac{\alpha^2+1}{\alpha^2-1}\right] \tag{2.4}$$

Constants of proportionality in Eq. (2.1) to Eq. (2.4) are demarcated by square brackets. These equations are valid when the size of the gap is small: $1 < \alpha \leq 1.10$. Wide gaps can lead to significant errors in determining rheological properties. Example Problem 8.3 examines the problem of finding rheological properties of ice cream mix from raw data obtained with a concentric cylinder viscometer.

2.3 Mixer Viscometry

Concentric cylinder viscometers are excellent for characterizing the rheological behavior of fluids for the purpose of pipeline design and pump selection. Some fluids, however, cannot be properly tested in narrow gap viscometers because they have one or more of the following characteristics:

- Particles too large for the gap causing added torque during testing. To avoid serious measurement errors the maximum cross-sectional dimension of the largest particles should be less than one-third the size of the gap: $(R_c - R_b)/3$. Since gap sizes on the order of 1 mm to 6 mm are typical, particle diameters are limited to a size of approximately 0.33 mm to 2 mm in concentric cylinder systems.
- Materials that undergo phase separation (a phenomenon called slip) during testing result in particle migration during bob rotation causing a more fluid layer to form at the testing surface.
- Gravity-induced particle settling producing a non-homogeneous sample during testing.

Mixer viscometers can often be used when any of the above problems make conventional rotational viscometers inappropriate.

The following sections provide practical information on mixer viscometry based on the k'' method (also called the torque curve method). In Sec. 2.3.1, **Characterizing the Mixer Viscometer**, the procedures to determine the mixer coefficient (k'') and the mixer viscometer constant (k') are presented. These techniques, requiring the use of numerous standard fluids, are needed if an untested mixer-impeller geometry is to be used for collecting rheological data. In Sec. 2.3.2, **Finding Properties of Power Law Fluids**, the procedure to calculate apparent viscosity as a function of k'' at an average shear rate defined by k' is given. This approach provides a straight forward means of generating a rheogram for a fluid with a mixer viscometer system that has already been characterized using the methods of Sec. 2.3.1.

2.3.1 Characterizing the Mixer Viscometer

Mixer viscometers must be properly characterized before they can be used to collect pipeline design data. The analysis is accomplished using dimensionless numbers and standard fluids having known rheological properties. Dimensional analysis of mixing Newtonian fluids in the laminar flow regime establishes the empirical relationship between the power number (N_{Po}) and the impeller Reynolds number ($N_{Re,I}$):

$$N_{Po} = \frac{A}{N_{\mathrm{Re},I}} \tag{2.5}$$

where:

$$N_{Po} = \frac{\Phi}{\rho \Omega^3 d^5} \tag{2.6}$$

$$N_{\mathrm{Re},I} = \frac{\rho \Omega d^2}{\mu} \tag{2.7}$$

A is a dimensionless constant. The variables Φ, d, and Ω are the power required to turn the impeller, the impeller diameter and angular velocity of the impeller, respectively. When the units of angular velocity are rad s^{-1}, flow is considered laminar if $N_{Re,I} < 63$.

Eq. (2.5) is used as the basis for determining the rheological properties of fluids. Substituting Eq. (2.6) and (2.7) into Eq. (2.5) gives

$$\frac{\Phi}{d^5 \Omega^3 \rho} = \frac{A\mu}{d^2 \Omega \rho} \tag{2.8}$$

Power (Φ) is defined as the product of the torque (M) and the angular velocity (Ω):

$$\Phi = M\Omega \tag{2.9}$$

Substituting Eq. (2.9) into Eq. (2.8), and simplifying the result yields an equation for the viscosity of a Newtonian fluid:

$$\mu = \frac{M}{A d^3 \Omega} = \frac{k'' M}{\Omega} \tag{2.10}$$

where k'' is a constant defined as the mixer coefficient: $k'' = A^{-1} d^{-3}$. Newtonian fluids with a known viscosity are used to determine k''. This constant ties together the dynamic relationship between the physical system (mixer, impeller, degree of fill), torque, angular velocity of the impeller, and viscosity.

To evaluate a power law fluid in a mixer viscometer, the apparent viscosity (η) is defined as a function of the average shear rate:

$$\eta = K\left(\dot{\gamma}_a\right)^{n-1} \tag{2.11}$$

Average shear rate ($\dot{\gamma}_a$) is defined as the product of the mixer viscometer constant (k') and the angular velocity:

$$\dot{\gamma}_a = k'\Omega \tag{2.12}$$

To determine the average shear rate, one must apply the matching viscosity assumption that the Newtonian viscosity and the apparent viscosity are equal at the same shear rate. Hence, Eq. (2.10) is set equal to Eq. (2.11):

$$\frac{k''M}{\Omega} = K\left(\dot{\gamma}_a\right)^{n-1} \tag{2.13}$$

Solving Eq. (2.13) for the average shear rate yields

$$\dot{\gamma}_a = \left(\frac{k''M}{\Omega K}\right)^{\frac{1}{n-1}} \tag{2.14}$$

The mixer coefficient (k'') and the mixer viscometer constant (k') can be found by executing the following steps:

1) Obtain Newtonian fluids and measure the viscosity at selected test temperatures using a rotational viscometer with conventional attachments such as the concentric cylinder. Candidate fluids include, honey, corn syrup, silicone oil and glycerin.
2) Using the "standard" Newtonian fluids evaluated from Step 1, collect experimental data of torque versus angular velocity in the mixer viscometer system. Using Eq. (2.10), written as

$$\mu\,\Omega = k''M \tag{2.15}$$

plot the product of viscosity and angular velocity (y-axis) versus torque (x-axis). Then, by regression analysis, determine k'' as the slope of Eq. (2.15). If the relationship is non-linear, the initial assumption that there is a linear relationship between the power number and the impeller Reynolds number (stated mathematically by Eq. (2.5)) is invalid, and k'' cannot be determined using this method. Experimental data taken with corn syrup and honey were used to establish k'' for a helical screw impeller in Example Problem 8.4.

3) Obtain power law fluids that can be tested in a conventional concentric cylinder viscometer, and determine the consistency coefficient and the flow behavior index of these materials at the appropriate temperatures. Candidate fluids include materials such as aqueous solutions of guar gum (1%), and methylcellulose (1.0, 2.0, and 2.5%).
4) Using the "standard" power law fluids evaluated from Step 3, collect experimental data of torque versus angular velocity in the mixer viscometer. Using these data, k'' found in Step 2, and Eq. (2.14), determine the average shear rate at each angular velocity.
5) Taking the data set of average shear rate versus angular velocity generated in Step 4, determine the mixer viscometer constant (using more regression analysis) as the slope of Eq. (2.12).

Although the principles of mixer viscometry are well established, commercial instrument companies have not aggressively applied it to established viscometer product lines. Some discussion of the published literature, current problems and opportunities in mixer viscometry is warranted.

Table 2.1. Typical values of k'' and k' for laboratory scale impellers.

Impeller	d*	D[#]	Gap[+]	k''	k'
	cm	cm	cm	rad m^{-3}	rad^{-1}
Helical Ribbon [1]	3.30	4.20	0.45	1,632	1.4
Interrupted Helical Screw [2]	5.30	9.30	2.00	3,047	1.6
Haake Pitched Paddle [3]	4.14	4.20	0.03	2,215	4.5
RVA Pitched Paddle [4]	3.28	3.66	0.19	12,570	3.2
Brookfield Flag Impeller [5]	1.50	1.90	0.20	61,220	2.9

* d = impeller diameter; [#]D = cup diameter; [+]gap = $(D-d)/2$

[1] Steffe, J.F., E. Agrawal, and K.D. Dolan. *J. Texture Stud.* (2003) 34: 41-52; [2] Omura, A.P. and J.F. Steffe. *J. Food Proc. Engr.* (2003) 26: 435-445; [3] Steffe, J.F. and E.W. Ford. *J. Texture Stud.* (1985) 16: 179-192; [4] Lai, K.P., J.F. Steffe and P.K.W. Ng. *Cereal Chem.* (2000) 77: 714-716; [5] Briggs, J.L. and J.F. Steffe. *J. Texture Stud.* (1996) 27: 671-677.

Typical values of the mixer coefficient constant and the mixer viscometer constant determined by the procedure given above are provided in Table 2.1. These constants are strictly empirical and only relate to the specific geometry (impeller and cup) combination tested. Other factors such as the fluid levels during testing, and spacing between the impeller and the cup, must also be maintained. One limitation of mixer viscometry is that low values of k' restrict the instruments to moderate shear rates that may be below the maximum shear rates found in pipelines. Also, the torque capacity of the instrument may limit the analysis to low shear rates when dealing with thick materials.

All mixers are useful in minimizing slip (phase separation) during testing. As a general guideline, maximum particle size should be limited to one-third the separation distance between the outer edge of the impeller and the container wall:

$$\text{maximum particle size} < \frac{D-d}{6} \qquad (2.16)$$

Fluids with particles exceeding this size recommendation may sometimes be successfully tested at low angular velocities. Also, when examining power law fluids with a mixer viscometer, laminar flow conditions should be maintained to generate acceptable data. Laminar flow is assumed if the power law impeller Reynolds number is less than 63:

$$N_{\mathrm{Re},Pl,I} = \frac{d^2\Omega\rho}{\eta} = \frac{d^2\Omega\rho}{K\left(k'\Omega\right)^{n-1}} < 63 \qquad (2.17)$$

with angular velocity (Ω) given in units of rad /s.

Illustrations of the sensors identified in Table 2.1 are provided in Fig. 2.3 – 2.7. Each system is unique:

- **Helical Ribbon (Fig. 2.3).** This impeller (manufactured in the Food Rheology Laboratory at Michigan State University, East Lansing, MI), was designed to study emulsions such as mayonnaise. During sensor rotation the fluid is simultaneously rotated and lifted on the ribbon before falling down the central axis to the bottom of the container.
- **Interrupted Helical Screw (Fig. 2.4).** This impeller (manufactured in the Food Rheology Laboratory at Michigan State University, East Lansing, MI) was constructed as a 1/3 scale replica of a successful fluid concrete testing system. The design promotes the movement of large particles. Fluids that have been successfully evaluated with this sensor include creamed corn, chunky salsa, and chunky tomato spaghetti sauce.

Figure 2.3. The helical ribbon impeller (diameter = 3.30 cm).

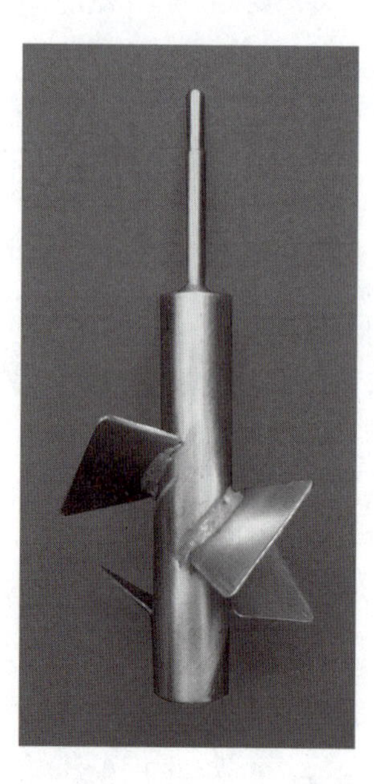

Figure 2.4. The interrupted helical screw impeller (diameter = 5.30 cm).

- **Haake Pitched Paddle (Fig. 2.5).** The impeller (manufactured by Thermo Electron Corporation, Waltham, MA) has two paddle blades pitched at 15 degrees with a 9 mm diameter hole drilled in them. This geometry keeps particles in suspension so gelatinization can be studied during the heating of aqueous starch solutions. It has also been used for various pureed foods.

- **RVA Pitched Paddle (Fig. 2.6).** The RVA impeller (Rapid Visco Analyser, Newport Scientific, Warriewood, Australia) is used to investigate the "cooked viscous properties of starch, grain, flour and foods." It characterizes the pasting behavior of aqueous starch solutions during heating and cooling. The pitched (and curved) blade maintains particle suspension during rotation.

- **Brookfield Flag Impeller (Fig. 2.7).** The Brookfield Flag Impeller (Brookfield Engineering Laboratories, Middleboro, MA) is intended for use in the Brookfield Small Sample Adapter (a 13 cc cup). This design has been successfully used for numerous food products.

2.3.2 Finding Properties of Power Law Fluids

Once the mixer viscometer system has been characterized by determining k'' and k', power law fluid properties can be determined from experimental data. The rotational speed of the instrument should be selected to match the shear rate range in the pipeline.

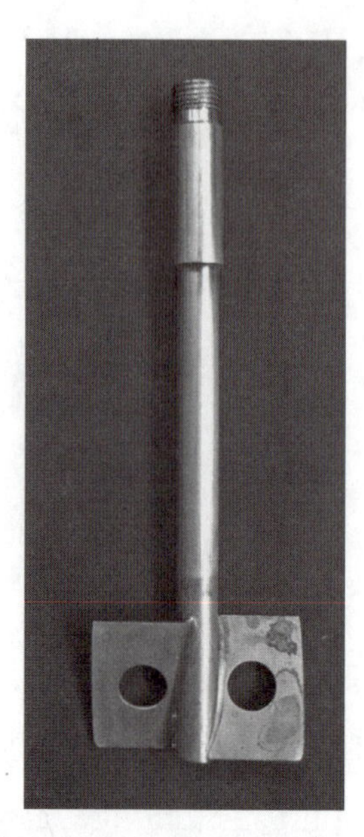

Figure 2.5 The Haake pitched (at an angle of 15 degrees) paddle impeller (diameter = 4.41 cm).

Figure. 2.6 The RVA pitched paddle impeller (diameter = 3.28 cm).

Figure 2.7. The Brookfield flag impeller (diameter = 1.50 cm).

This is accomplished using k': Since $\Omega_{min} = \dot{\gamma}_{min} / k'$ and $\Omega_{max} = \dot{\gamma}_{max} / k'$, then the best range for Ω is

$$\frac{\dot{\gamma}_{min}}{k'} \leq \Omega \leq \frac{\dot{\gamma}_{max}}{k'} \tag{2.18}$$

Once the instrument rotational speed range is established, a series of data points of torque (M) versus angular velocity (Ω) is collected. The matching viscosity assumption is applied using k'' so the apparent viscosity can be calculated at each speed:

$$\eta = \frac{M\,k''}{\Omega} \tag{2.19}$$

The corresponding average shear rate, defined by Eq. (2.12), is also calculated at each speed. Resulting values of apparent viscosity versus average shear rate are plotted and, by regression analysis of these data using Eq. (2.11), the flow behavior index and the consistency coefficient are found. This procedure is applied to pasta sauce with tomato and mushroom pieces in Example Problem 8.5.

2.3.3 Brookfield Viscometers with Disk Spindles

Brookfield viscometers (Brookfield Engineering Laboratories, Middleboro, MA) equipped with flat disk shaped spindles (Fig. 2.8) are often used to examine the behavior of biological fluids. Unfortunately, the shear rate over a flat disk is non-uniform and difficult to describe mathematically. However, by taking a mixer viscometry approach, and defining an average shear stress and an average shear rate, these sensors can be used to determine non-Newtonian fluid properties. The method presented here was developed by Mitschka [Rheol. Acta (1982)12: 207-209], and extended by Briggs and Steffe [J. Texture Studies (1997) 28: 517-522].

The first step is to determine the flow behavior index (n) which can be can be found from the following equation:

$$M = (\text{constant})\, N^n \tag{2.20}$$

or, in the logarithmic form, as

$$\ln M = \ln(\text{constant}) + n \, \ln(N) \tag{2.21}$$

Ω (radians per second) can be used in place of N (revolutions per minute) in Eq. (2.20) and Eq. (2.21) with no change in the calculated value of n. N is used here for convenience because Brookfield operators select speed in revolutions per minute when running the instrument. Also, note that percent torque, the typical parameter obtained from the Brookfield instrument, can be substituted for M in the above equations with no loss of accuracy in finding the flow behavior index.

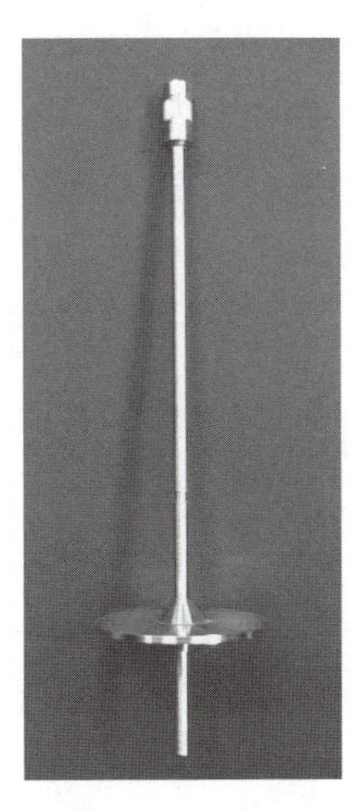

Figure 2.8. Brookfield disk type spindle #3 (diameter = 3.50 cm).

The average shear stress is calculated as

$$\sigma_a = k_\sigma (C)(\text{percent torque}) \tag{2.22}$$

where k_σ, known as the shear stress conversion factor, changes with the spindle number (Table 2.2). The value of C (a dimensionless constant) depends on the total torque capacity of the instrument (Table 2.3).

Table 2.2. Shear stress conversion factors (k_σ) for Brookfield spindles.

Brookfield Spindle Number	k_σ, Pa
1	0.035
2	0.119
3	0.279
4	0.539
5	1.05
6	2.35
7	8.40

Table 2.3. Values of *C* for typical Brookfield viscometers.

Viscometer Model	Max. Torque, dyne cm	*C*, dimensionless
½ RV	3,584	0.5
RV	7,187	1.0
HAT	14,374	2.0
HBT	57,496	8.0

The "percent torque" is the percentage of the maximum torque recorded during testing at a constant speed. Average shear rate is

$$\dot{\gamma}_a = k_{\dot{\gamma}}\left(N\right) \tag{2.23}$$

where $k_{\dot{\gamma}}$, known as the shear rate conversion factor, depends on the numerical value of the flow behavior index:

$$k_{\dot{\gamma}} = 0.263\left(\frac{1}{n}\right)^{0.771} \tag{2.24}$$

Standard Brookfield sample handling methods are followed to determine rheological properties. Samples are placed in a 600 ml low form Griffin beaker, and a spindle is inserted into the material up to the notch on the shaft. After a series of data points of torque versus angular velocity (rpm) are collected, the data analysis procedure described here is used to determine rheological properties.

The flow behavior index (n) is found from Eq. (2.21). Average shear stress and the average shear rate are calculated using Eq. (2.22) and Eq. (2.23), respectively. Power law fluid properties are determined by regression analysis of the resulting values. The flow behavior index determined in this step will be identical to the value found previously using Eq. (2.21).

2.4 Cone and Plate, and Parallel Plate Viscometers

Cone and plate sensors are commonly made available with rotational viscometers. Cone angles (Fig. 2.9) are very small, typically 2 to 4 degrees. The advantage of the cone and plate geometry is that the shear stress and shear rate are uniform throughout the gap, and may be easily calculated as the angular velocity divided by the tangent of the cone angle, θ:

$$\dot{\gamma} = \frac{\Omega}{\tan\theta} \tag{2.25}$$

The corresponding shear stress is directly proportional to the torque and inversely proportional to the radius cubed:

$$\sigma = \frac{3M}{2\pi R^3} \tag{2.26}$$

In a parallel plate system (Fig. 2.9), the shear rate is calculated at the outer rim of the plate as a function of the angular velocity, the radius, and the distance separating the plates (h):

$$\dot{\gamma}_R = \frac{\Omega R}{h} \tag{2.27}$$

The value of h is typically 0.5 to 2.0 mm. Calculating the shear stress at the rim may require a correction for non-Newtonian phenomena. Assuming power law behavior, it can be estimated as

$$\sigma_R = \frac{M(3+n)}{2\pi R^3} \tag{2.28}$$

where

$$n = \frac{d\ln(M)}{d\ln(\dot{\gamma}_R)} = \frac{d\ln(M)}{d\ln(\Omega)} \tag{2.29}$$

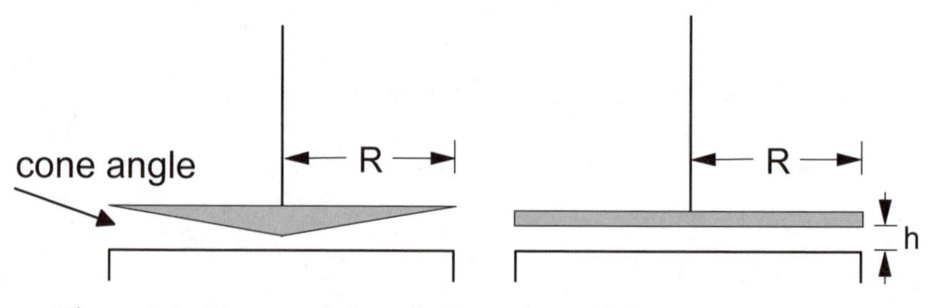

Figure 2.9. Cone and plate (left), and parallel plate (right) sensors.

3 Energy and Head Loss Calculations

3.1 Mechanical Energy Balance Equation

Pump selection requires an evaluation of the energy and power required to move fluid through a pipeline. This transport is evaluated using the overall mechanical energy balance equation (sometimes called the modified Bernoulli equation) that may be written as

$$\frac{P_1}{\rho}+\frac{\overline{u}_1^2}{\alpha}+gz_1+W=\frac{P_2}{\rho}+\frac{\overline{u}_2^2}{\alpha}+gz_2+\sum F \tag{3.1}$$

or, when solved for W (the work input from pump, J/ kg) as

$$W=\frac{P_2-P_1}{\rho}+\frac{(\overline{u}_2^2-\overline{u}_1^2)}{\alpha}+g(z_2-z_1)+\sum F \tag{3.2}$$

Each term in Eq. (3.2) represents the energy change (which may be negative or positive) per unit mass:

$$\frac{P_2-P_1}{\rho}=\text{pressure energy loss, J/kg} \tag{3.3}$$

$$\frac{(\overline{u}_2^2-\overline{u}_1^2)}{\alpha}=\text{kinetic energy loss, J/kg} \tag{3.4}$$

$$g(z_2-z_1)=\text{potential energy loss, J/kg} \tag{3.5}$$

$$\sum F=\text{summation of all friction losses, J/kg} \tag{3.6}$$

Subscripts one and two refer to two different locations within the control volume of the pumping system. The mechanical energy balance equation is applied to pipeline design challenges in Example Problems 8.7 and 8.8.

Values of the kinetic energy correction factor (α) may be determined as follows: $\alpha = 2.0$ for turbulent flow of any fluid; $\alpha = 1$ for laminar flow of Newtonian fluids; and for laminar flow of power law fluids, α is a function of the flow behavior index:

$$\alpha = \frac{2(2n+1)(5n+3)}{3(3n+1)^2} \tag{3.7}$$

The summation of friction losses described by Eq. (3.6) includes all of the following:

1) Losses in straight pipe with a separate term for each diameter:

$$F = \frac{2 f \bar{u}^2 L}{D} \tag{3.8}$$

where f is the Fanning friction factor. Procedures for determining f are presented in Chapter 4. For noncircular cross sections, the hydraulic diameter (D_h) may be used in place of the inside tube diameter:

$$D_h = \frac{4(\text{flow area of cross section})}{\text{wetted perimeter}} \tag{3.9}$$

In the case of flow through an annulus, a common configuration involving flow between two concentric tubes, the hydraulic diameter is

$$D_h = D_{outer} - D_{inner} \tag{3.10}$$

where D_{outer} and D_{inner} are, respectively, the outer and inner diameters defining annular flow area.

2) Losses in valves and fittings calculated using the velocity head method (the preferred method in this text):

$$F = \frac{k_f \bar{u}^2}{2} \tag{3.11}$$

or the equivalent length (L_e) method:

$$F = \frac{2 f \bar{u}^2 L_e}{D} \tag{3.12}$$

Procedures for determining k_f are presented in Chapter 5.

3) Losses in other equipment, such as flow meters and pneumatic valves, are added using data provided by the manufacturer as pressure drop across the equipment (when operated at the required flow rate for water) divided by the density:

$$F = \frac{\Delta P}{\rho} \tag{3.13}$$

A method to account for energy losses for other fluids, based on data for water, is discussed in Sec. 5.2. The method is applied to a Newtonian fluid (cream) in Example Problem 8.7, and to a non-Newtonian fluid (pulpy fruit juice) in Example Problem 8.8.

Hydraulic power (Φ) is defined as the work input times the mass flow rate:

$$\Phi = W \dot{m} \tag{3.14}$$

and pump efficiency (not to be confused with apparent viscosity that is defined with the same symbol, η) is

$$\eta = \frac{\text{power imparted to the fluid by the pump}}{\text{power input to the pump}} = \frac{\text{hydraulic power}}{\text{required power}} \tag{3.15}$$

The required power, also called the brake horsepower, represents the input power needed at the pump shaft. It includes the power required to transport fluid through the system (the hydraulic power), and the power required to overcome friction losses in the pump itself.

3.2 System Curves (Process Requirements)

System curves describe the energy input requirements needed to achieve specific processing objectives. They are obtained by solving the mechanical energy balance equation in terms of head (given in units of m). Shaft work (W) can be defined using the total system head (H_s):

$$W = H_s g \tag{3.16}$$

H_s is the total discharge head minus the total suction head. Combining the definition of shaft work given by Eq. (3.2) and Eq. (3.16) gives H_s in terms of the pressure, velocity, elevation, and friction heads:

$$H_s = \frac{(P_2 - P_1)}{\rho g} + \frac{(\bar{u}_2^2 - \bar{u}_1^2)}{\alpha g} + (z_2 - z_1) + \frac{\sum F}{g} \tag{3.17}$$

Each collection of terms in Eq. (3.17) represents the change (negative or positive) in head:

$$\frac{(P_2 - P_1)}{\rho g} = \text{pressure head, m} \tag{3.18}$$

$$\frac{(\bar{u}_2^2 - \bar{u}_1^2)}{\alpha g} = \text{velocity head, m} \tag{3.19}$$

$$(z_2 - z_1) = \text{elevation head, m} \tag{3.20}$$

$$\frac{\sum F}{g} = \text{friction head, m} \tag{3.21}$$

Although head is expressed in units of m, it actually represents energy per unit weight. Dividing the changes in energy per unit mass, expressed by Eq. (3.3) to Eq. (3.6), by g results in units of Joules per Newton. In the SI system, a Joule per Newton has the unit of length: meter.

H_s is a function of the volumetric flow rate expressed in units of m^3s^{-1} or gpm. Plotting the total system head versus volumetric flow rate gives the system curve required for the process. The velocity head (accounting for kinetic energy differences) is usually small and often ignored in head calculations. System curves are required to evaluate centrifugal pumps for specific applications. They are not required to size positive displacement pumps because flow rate is largely independent of the friction losses found in the process. A system curve for cream is generated in Example Problem 8.7, and calculations for a positive displacement pump are given in Example Problem 8.8

3.3 Finding Work and Head

Investigating some typical applications is a useful way to examine the mechanical energy balance equation. A very common problem involves pumping fluid from a tanker truck to a storage tank. Flexible pipe from the truck attaches to rigid pipe near the pump. Assume the tanks have open hatches so the pressure is one atmosphere at the upper fluid level in each tank. Fig. 3.1 illustrates an unloading situation where the pressure at points one and two are equal: $P_1 = P_2$. Since the movement of the liquid surfaces is very slow, one can assume that velocities at points one and two are essentially zero: $\bar{u}_1 = \bar{u}_2 = 0$. The energy loss in the strainer is calculated in terms of the pressure drop across the strainer (at the specified flow rate) divided by the fluid density: $\Delta P/\rho$. This information is based on experimental data for water provided by the company manufacturing the strainer. It can be adjusted for non-Newtonian fluids using the method given in Sec. 5.2.

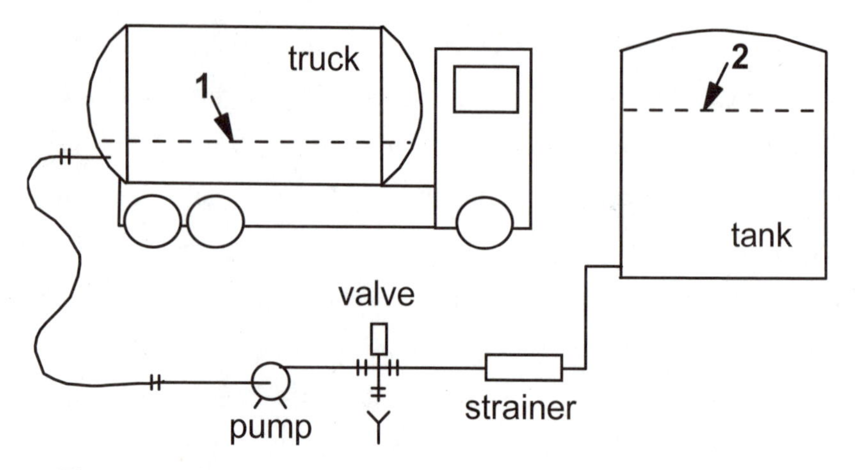

Figure 3.1. Process line between tanker truck and storage tank with control volume located between fluid levels.

Energy loss in both the rigid and slightly curved flexible sections of pipe (Fig. 3.1) is calculated using the Fanning friction factor. Losses in the entrance from the tanker to the pipe, two elbows, the valve, and the exit from the pipe to the storage tank are evaluated using friction loss coefficients. Additional terms are required if different tube sizes are used in the processing line because this change affects $\overline{u}$, and f. Also note that the largest energy demands on the pump occur when the tanker is nearly empty and the storage tank is nearly full because at that point the pump is working against the maximum elevation head $(z_2 - z_1)$. Analytical results (Table 3.1) may be evaluated in terms of work (W) or the total system head ($H_s = W / g$).

The problem illustrated in Fig. 3.2 is very similar to the problem shown in Fig. 3.1, except point two is selected just prior to the discharge location of the pipe. In this application of the mechanical energy balance equation, a number of new considerations must be introduced: the velocity at the exit of the control volume is not zero because the fluid is still in the pipe; P_1 = 1 atm (absolute pressure) but P_2 > 1 atm (absolute pressure) and must be retained in the solution; the friction loss at the pipe exit is irrelevant because it is beyond the control volume; friction losses include three elbows. Other elements of the equation remain unchanged and the solution for shaft work is given in Table 3.1.

Table 3.1. Applications of the mechanical energy balance equation applied to Figures 3.1 to 3.6.

Shaft work and total system head for the problem illustrated in Figure 3.1 $W = g(z_2 - z_1) + \frac{\bar{u}^2}{2}\left(k_{f,\text{ entrance}} + 2k_{f,\text{ 90°elbow}} + k_{f,\text{ valve}} + k_{f,\text{ exit}}\right) + \frac{2f\bar{u}^2L}{D} + \left(\frac{\Delta P}{\rho}\right)_{\text{strainer}}$ $H_s = (z_2 - z_1) + \frac{\bar{u}^2}{2g}\left(k_{f,\text{ entrance}} + 2k_{f,\text{ 90°elbow}} + k_{f,\text{ valve}} + k_{f,\text{ exit}}\right) + \frac{2f\bar{u}^2L}{Dg} + \left(\frac{\Delta P}{\rho g}\right)_{\text{strainer}}$
Shaft work for the problem illustrated in Figure 3.2 $W = \frac{P_2 - P_1}{\rho} + g(z_2 - z_1) + \frac{\bar{u}_2^2}{\alpha} + \frac{\bar{u}^2}{2}\left(k_{f,\text{ entrance}} + 3k_{f,\text{ 90°elbow}} + k_{f,\text{ valve}}\right) + \frac{2f\bar{u}^2L}{D} + \left(\frac{\Delta P}{\rho}\right)_{\text{strainer}}$
Shaft work for the problem illustrated in Figure 3.3 $W = g(z_2 - z_1) + \frac{\bar{u}^2}{2}\left(k_{f,\text{ entrance}} + 3k_{f,\text{ 90°elbow}} + k_{f,\text{ valve}} + k_{f,\text{ exit}}\right) + \frac{2f\bar{u}^2L}{D} + \left(\frac{\Delta P}{\rho}\right)_{\text{strainer}}$
Pump pressure differential for the problem illustrated in Figure 3.4 $P_2 - P_1 = W\rho$
Pressure at the pump exit *(P_1)* for the problem illustrated in Figure 3.5 $P_1 = P_2 + \rho g(z_2 - z_1) - \frac{\rho\bar{u}_1^2}{\alpha} + \frac{\rho\bar{u}^2}{2}\left(2k_{f,\text{ 90°elbow}} + k_{f,valve} + k_{f,\text{ exit}}\right) + \frac{2\rho f\bar{u}^2L}{D} + (\Delta P)_{\text{strainer}}$
Pressure at pump entrance *(P_2)* for the problem illustrated in Figure 3.6 $P_2 = P_1 + \rho g(z_1 - z_2) - \frac{\rho\bar{u}_2^2}{\alpha} - \frac{\rho\bar{u}^2}{2}\left(k_{f,\text{ entrance}}\right) - \frac{2\rho f\bar{u}^2L}{D}$

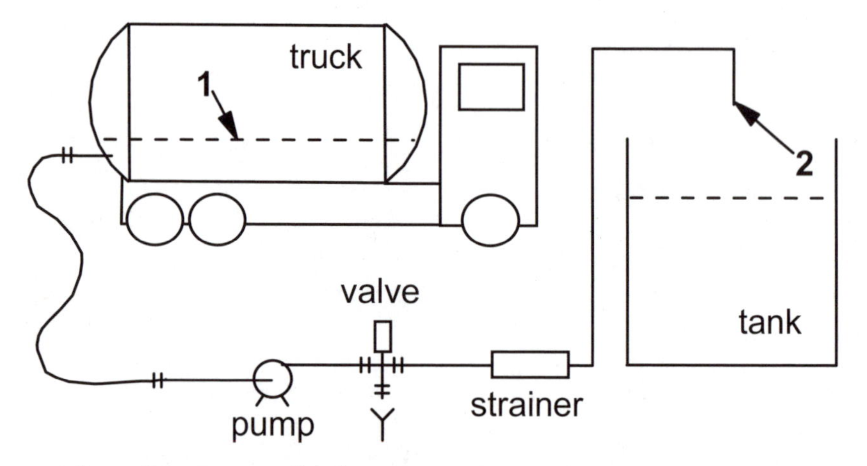

Figure 3.2. Process line between tanker truck and storage tank with control volume located between the fluid level in the truck and a point just inside the pipe exit.

A related problem is illustrated in Fig. 3.3. In this case, however, the control volume is selected at a point outside the pipe where the surface of the fluid reaches a pressure of one atmosphere and the kinetic energy has largely dissipated. This situation requires that the friction loss in the exit be included in the calculation, but allows elimination of pressure energy because $P_1 = P_2$. Remaining parts of the solution provided for Fig. 3.1 are unchanged (Table 3.1). The mechanical energy balance equation [Eq. (3.2)] can also be applied across the pump itself (Fig. 3.4). This is useful in predicting the pressure differential across the pump given a known value of W (Table 3.1).

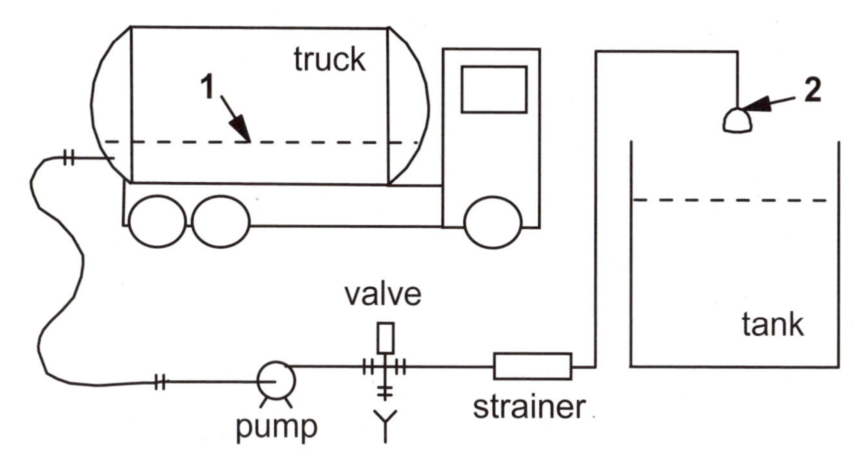

Figure 3.3. Process line between tanker truck and storage tank with control volume located between the fluid level in the truck and a point just outside the pipe exit.

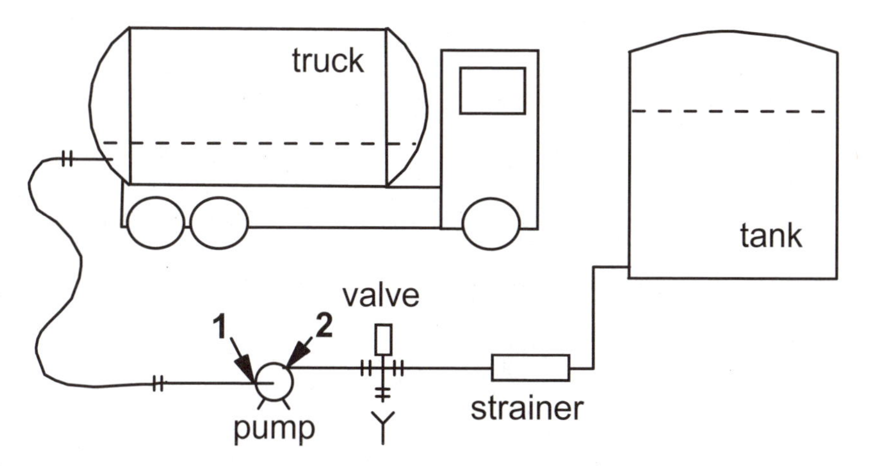

Figure 3.4. Process line between tanker truck and storage tank with control volume located between the inlet and outlet of the pump.

Another practical problem is illustrated in Fig. 3.5 where the control volume is defined between the exit of the pump and the exit to the tank. In this case, $W = 0$ (there is no pump in the control volume), $\bar{u}_2 = 0$ (pumping into a large tank), and friction losses occur in the valve, the strainer, two elbows, the exit, and the straight pipe. Given these factors, the actual pressure at the pump exit may be calculated (Table 3.1). The information may be needed to evaluate seal integrity for a pump. Fig. 3.6 illustrates the analogous problem of finding the pressure at the pump inlet. The solution is given in Table 3.1. This problem must be solved to determine the net positive suction head available in a pumping system – more on this in Sec. 3.5.

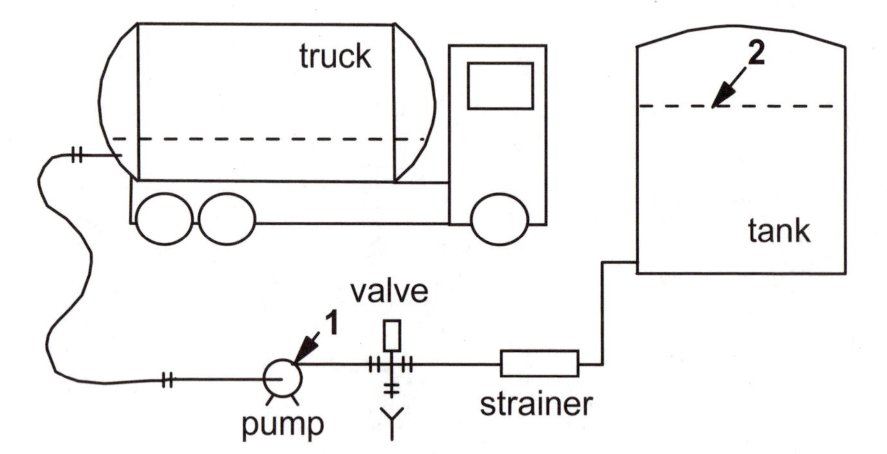

Figure 3.5. Process line between tanker truck and storage tank with control volume located between the pump outlet and the system exit.

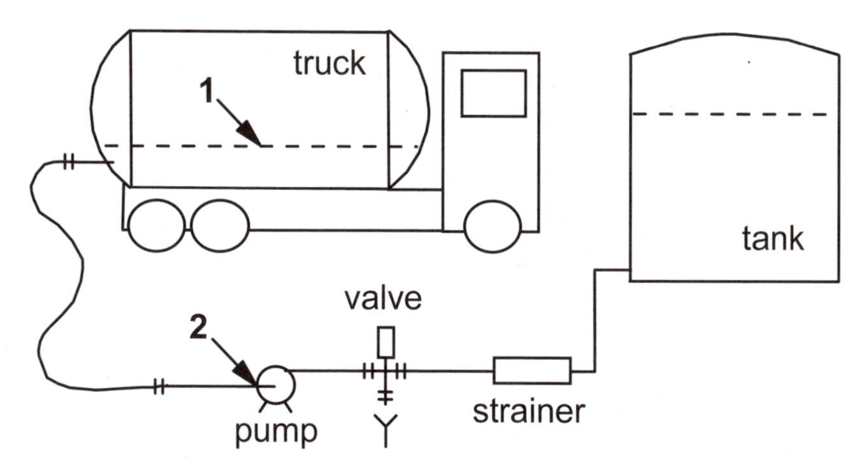

Figure 3.6. Process line between tanker truck and storage tank with control volume located between the tanker truck and the pump entrance.

3.4 Pump Curves (Centrifugal Pumps)

System curves must be compared to pump curves to determine the actual conditions of operation for centrifugal pumps. These curves are determined for a particular pump, operating at a constant speed (typically 1750 or 3500 rpm), by measuring the pressure at the inlet (P_1) and the outlet (P_2) while moving different volumes of water (m^3s^{-1} or gpm) through the equipment. Resulting data can be used to determine the head generated by the pump called the total pump head (H_p):

$$H_p = \frac{(P_2 - P_1)}{\rho g} \tag{3.22}$$

The intersection of the two plots (H_s vs. Flow Rate, and H_p vs. Flow Rate) gives the operational flow rate of a centrifugal pump (Fig. 3.7). Pump curves are not used to evaluate the performance of positive displacement (PD) pumps because the flow rate in a PD pump is more or less constant regardless of the pressure.

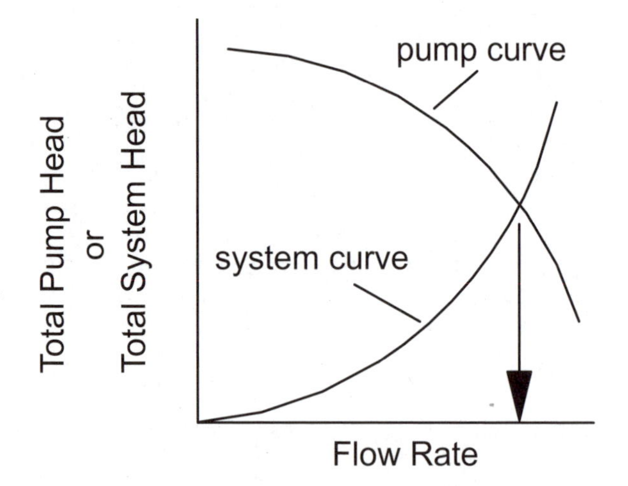

Figure 3.7. Intersection of pump (H_p) and system (H_s) curves showing the flow rate through a centrifugal pump.

3.5 Net Positive Suction Head (Available and Required)

To avoid cavitation problems, the net positive suction head available must be greater than the net positive suction head required: $(NPSH)_A > (NPSH)_R$. The $(NPSH)_R$ is taken from experimental data for a particular pump. To determine the $(NPSH)_A$, the mechanical energy balance equation [Eq. (3.1)] must be solved for the inlet pressure at the pump. This problem is illustrated in Fig. 3.6 where point 1 is at the fluid level of the inlet tank and point 2 is located at the pump inlet. Evaluating the mechanical energy balance equation leads to the following general solution for the inlet pressure at the pump:

$$P_2 = P_1 + \frac{\rho\left(\overline{u}_1^2 - \overline{u}_2^2\right)}{\alpha} + \rho g\left(z_1 - z_2\right) - \rho\sum F \qquad (3.23)$$

If the pump inlet pressure is less than the vapor pressure of the liquid, vapor bubbles can form causing the phenomenon known as cavitation. Eq. (3.23) provides a mathematical summary of the circumstances that may cause cavitation: 1) pumping from an evacuated chamber making P_1 very low; 2) significantly increasing the velocity of the fluid during pumping making the velocity at point one much greater than the velocity at point two; 3) requiring the pump to substantially elevate the liquid making z_2 much larger than z_1; 4) having a large friction loss in the inlet line to the pump caused by an excessive length of pipe, or too many valves and fittings, or other flow obstructions such as heat exchangers and flow meters.

To evaluate $(NPSH)_A$, both the absolute pressure at the pump (P_2) and the vapor pressure (P_v) of the fluid must be considered. This is accomplished by subtracting the vapor pressure from the total pressure (to be sure there is adequate pressure to avoid cavitation), and dividing by $g\rho$ to get the results in units of head, m. Thus, the net positive suction head available is calculated as

$$(NPSH)_A = \frac{P_2}{g\rho} - \frac{P_v}{g\rho} = \frac{P_2 - P_v}{g\rho} \tag{3.24}$$

Values of the saturation vapor pressure of water, the usual solvent for biological fluids, are provided in Appendix 9.5.

Substituting the solution for P_2, given by Eq. (3.23), into Eq. (3.24) gives an equation for the net positive suction head available in terms of the recognized components of the mechanical energy balance:

$$(NPSH)_A = \frac{P_1}{g\rho} + \frac{\left(\overline{u}_1^2 - \overline{u}_2^2\right)}{\alpha g} + \left(z_1 - z_2\right) - \frac{\sum F}{g} - \frac{P_v}{g\rho} \tag{3.25}$$

where:

$\frac{P_1}{g\rho} =$ pressure head on the surface of the supply liquid, m. P_1 is equal to atmospheric pressure if the fluid supply is from an open tank.

$\frac{\left(\overline{u}_1^2 - \overline{u}_2^2\right)}{\alpha g} =$ velocity head, m. $\overline{u}_1$ is assumed zero if pumping from a large tank, and $\overline{u}_2$ is the average velocity at the pump inlet.

$\left(z_1 - z_2\right) =$ elevation head above the centerline of the tank, m. The physical situation may be described as suction lift if z_2 is greater than z_1, or positive (flooded) suction if z_1 is greater than z_2. Figure 3.6 illustrates an example of positive suction.

$\frac{\sum F}{g} =$ suction head loss due to the friction in piping, valves, fittings, etc., m.

$\frac{P_v}{g\rho} =$ vapor pressure head of the fluid being pumped, m. Vapor pressure for water, and most biofluids containing significant amounts of water, may be obtained from the properties of water given on saturated steam tables.

All pressures used in the above calculations are absolute. Recall that absolute pressure is gage pressure plus atmospheric pressure.

4 Fanning Friction Factors

4.1 Friction Factors

Friction factors are dimensionless numbers needed to determine energy losses in pipelines. Two friction factors – the Fanning friction factor and the Darcy friction factor – are commonly found in published literature. The Darcy value is equal to four times the Fanning value. Both produce identical results when used in approximate engineering calculations. Methods for calculating the Fanning friction factor (f) for Newtonian and power law fluids are presented in this chapter.

Surface roughness (ε) characterizes the surface condition of a pipe. The 3-A sanitary standard [3-A Sanitary Standards, Inc., McLean, Virginia, USA] for polished metal tubing for milk and milk products (3-A Standard Number 33-01) specifies a stainless steel surface finish equivalent to 150 grit resulting in a maximum roughness of 32 micro-inch (0.8 μm). A 180 grit finish may be required for pharmaceutical pipelines. Surface roughness for stainless steel pipe with commercial finishes is very small (Table 4.1). Polished stainless steel tubing may be considered smooth for the purpose of calculating Fanning friction factors at low Reynolds numbers. Small differences may be observed at large Reynolds numbers.

4.2 Newtonian Fluids

Newtonian fluids are defined as materials having a linear relationship between shear stress and shear rate: $\sigma = \mu \dot{\gamma}$. The Reynolds Number describing these materials in tube flow is

$$N_{\mathrm{Re}} = \frac{D\bar{u}\rho}{\mu} \tag{4.1}$$

where the volumetric average velocity ($\bar{u}$) is calculated as the volumetric flow rate (Q) divided by the cross-sectional area (A) of the pipe:

$$\bar{u} = \frac{V}{A} = \frac{4Q}{\pi D^2} \tag{4.2}$$

Table 4.1. Surface roughness (ε) of various pipeline materials.

Material	*ε, micro-inches*	ε, μm
commercial steel	1970	50
drawn copper/brass	59	1.5
stainless steel: 120 grit finish	43-48	1.09 -1.22
stainless steel: 150 grit finish	30-35	0.76 – 0.89
stainless steel: 180 grit finish	23-28	0.58 – 0.71
stainless steel: 240 grit finish	15-20	0.38 – 0.51
stainless steel: 320 grit finish	9-13	0.23 – 0.33
smooth tubing	0	0

The Fanning friction factor in laminar flow (defined as $N_{\mathrm{Re}} < 2100$) is calculated as

$$f_{\mathrm{laminar}} = \frac{16}{N_{\mathrm{Re}}} \tag{4.3}$$

The Blasius equation can be used to predict the Fanning friction factor for smooth tubes for turbulent flow N_{Re} numbers from 5000 to 100,000:

$$f_{turbulent} = \frac{0.0791}{N_{\mathrm{Re}}^{\;0.25}} \tag{4.4}$$

Haaland [J. Fluids Eng. March, 1983, pg 89-90] provided an equation to predict the friction factor in turbulent flow ($N_{Re} > 4100$) that included the effect of relative roughness (ε/D):

$$f = \left(\frac{1}{4}\right)\left\{-0.782\ \ln\left[\frac{6.9}{N_{\mathrm{Re}}} + \left(\frac{\varepsilon}{3.7D}\right)^{1.11}\right]\right\}^{-2} \tag{4.5}$$

The following equation published by Churchill [Chem. Engr. Nov 7, 1977, pg 91], covers all flow regimes (laminar, transition, and turbulent) and also includes the roughness effect:

$$f = 2\left[\left(\frac{8}{N_{\mathrm{Re}}}\right)^{12} + \frac{1}{\left(A+B\right)^{3/2}}\right]^{1/12} \tag{4.6}$$

where:

$$A = \left[2.457\ \ln\left(\frac{1}{\left(7/N_{\mathrm{Re}}\right)^{0.9} + 0.27\varepsilon/D}\right)\right]^{16} \tag{4.7}$$

$$B = \left(\frac{37{,}530}{N_{\mathrm{Re}}}\right)^{16} \tag{4.8}$$

The Fanning friction factor of a Newtonian fluid (cream) is calculated in Example Problem 8.7.

4.3 Power Law Fluids

Power law fluids have a non-linear relationship between shear stress and shear rate: $\sigma = K\dot{\gamma}^{n}$. Types of power law fluids are characterized by the numerical value of the flow behavior index: $0 < n < 1$

for shear-thinning fluids (very common), $n = 1$ for the special case of a Newtonian fluid, and $n > 1$ for shear-thickening fluids (uncommon). The power law Reynolds Number for tube flow is

$$N_{\mathrm{Re},PL} = \left(\frac{D^n \left(\overline{u} \right)^{2-n} \rho}{8^{n-1} K} \right) \left(\frac{4n}{3n+1} \right)^n \tag{4.9}$$

or, in another form commonly found in published literature, as

$$N_{\mathrm{Re},PL} = \frac{2^{3-n} D^n \overline{u}^{2-n} \rho}{K \left[\frac{(3n+1)}{n} \right]^n} \tag{4.10}$$

The term "generalized Reynolds number" is a synonym for power law Reynolds number. Using Eq. (4.9) versus Eq. (4.10) is a matter of personal preference because both equations produce the same numerical result.

Laminar flow of a power law fluid exists in the tube when

$$N_{\mathrm{Re},PL} < \left(N_{\mathrm{Re},PL} \right)_{critical} \tag{4.11}$$

The critical value of the power law Reynolds number depends on the numerical value of the flow behavior index [Darby, R. 2001. Chemical Engineering, March, pg 66-73]:

$$\left(N_{\mathrm{Re},PL} \right)_{critical} = 2100 + 875(1-n) \tag{4.12}$$

Values of the critical Reynolds number vary (Table 4.2) from 2888 at $n = 0.1$ to the familiar value of 2100 for Newtonian fluids when $n = 1.0$.

In the laminar flow regime, the Fanning friction factor is calculated with an equation analogous to Eq. (4.3):

$$f_{\text{laminar}} = \frac{16}{N_{\text{Re},PL}} \tag{4.13}$$

Experimental data indicate that laminar flow friction factors for foods may be somewhat lower. Work by Rozema and Beverloo [Lebensm-Wiss u. Technol. (1974) 7: 222-228] with biological fluids (mostly foods) gave the following results:

$$f_{\text{laminar}} = \frac{C}{N_{\text{Re},PL}} \tag{4.14}$$

where the average value of C was 14.3 (standard deviation = 3.1). Steffe et al. [Trans. ASAE (1984) 27: 616-619] found similar C values for power law fluid foods.

Table 4.2. Critical power law Reynolds numbers at different flow behavior indexes.

n	$\left(N_{\text{Re},PL}\right)_{critical}$
0.1	2888
0.2	2800
0.3	2713
0.4	2625
0.5	2538
0.6	2450
0.7	2363
0.8	2275
0.9	2188
1.0	2100

Fanning friction factors in the turbulent flow regime ($N_{Re,PL}$ > 4100) for smooth tubes can be calculated with the equation provided by Dodge and Metzner [AIChE J. (1959) 5: 189-204)]:

$$\frac{1}{\left(f_{turbulent}\right)^{1/2}}=\left(\frac{4}{n^{0.75}}\right)\log_{10}\left[N_{\mathrm{Re},PL}\,f_{turbulent}{}^{1-n/2}\right]-\left(\frac{0.4}{n^{1.2}}\right) \quad (4.15)$$

The influence of surface roughness on the value of the friction factor for power law fluids in turbulent flow is unknown; hence, omitted from the equation. Fortunately, this is not a problem in bioprocessing systems where stainless steel pipes, having a very low surface roughness, are the preferred material. Solving Eq. (4.15) for f when $N_{Re,PL}$ is known is cumbersome because the equation is implicit in f.

An alternative to Eq. (4.15) that is explicit in f, is the following relationship [Darby, R., Mun, R., Boger, V.B. 1992. Chemical Engineering, September, pg 116-119]. It can be used to predict the Fanning friction factor for power law fluids in smooth pipe over the entire range of power law Reynolds numbers:

$$f=(1-\alpha)f_{\mathrm{laminar}}+\frac{\alpha}{\left[\left(f_{turbulent}\right)^{-8}+\left(f_{transition}\right)^{-8}\right]^{1/8}} \quad (4.16)$$

where:

$$\alpha=\frac{1}{1+4^{-\Delta}} \qquad (\alpha=0 \text{ or } 1) \quad (4.17)$$

$$\Delta=N_{\mathrm{Re},PL}-\left(N_{\mathrm{Re},PL}\right)_{critical} \quad (4.18)$$

$$f_{\text{laminar}} = \frac{16}{N_{\text{Re},PL}} \tag{4.19}$$

$$f_{\text{turbulent}} = \frac{0.0682\, n^{-.5}}{N_{\text{Re},PL}^{\;1/(1.87+2.39n)}} \tag{4.20}$$

$$f_{\text{transition}} = 1.79 x 10^{-4} \left(\exp\left(-5.24n\right)\right)\left(N_{\text{Re},PL}^{\;0.414+.757n}\right) \tag{4.21}$$

Plotting the above equations clearly indicates that the flow behavior index can have a very significant impact on the friction factor. Graphical values of the Fanning friction factor based on Eq. (4.16) to Eq. (4.21) are given in Fig. 4.1 and 4.2; and values in tabular form are provided in Appendix 9.11. Curves overlap at low values of $N_{\text{Re},Pl}$ (Fig. 4.1). The Fanning friction factor for a power law fluid (pulpy fruit juice) is calculated in Example Problem 8.8.

4.4 Tube Flow Velocity Profiles

Continuous thermal processing systems for biological fluids include a heat exchanger (to heat the fluid); a length of tube known as the "hold tube" where the fluid residence time is sufficient to have an adequate thermal treatment; and usually includes a second heat exchanger (to cool the fluid). Because the hold tube is such a critical component of the system, understanding velocity profiles found in tube flow is important for numerically characterizing a thermal process. Velocity profiles in turbulent flow are given as a function of the Fanning friction factor; hence, they are presented in this chapter.

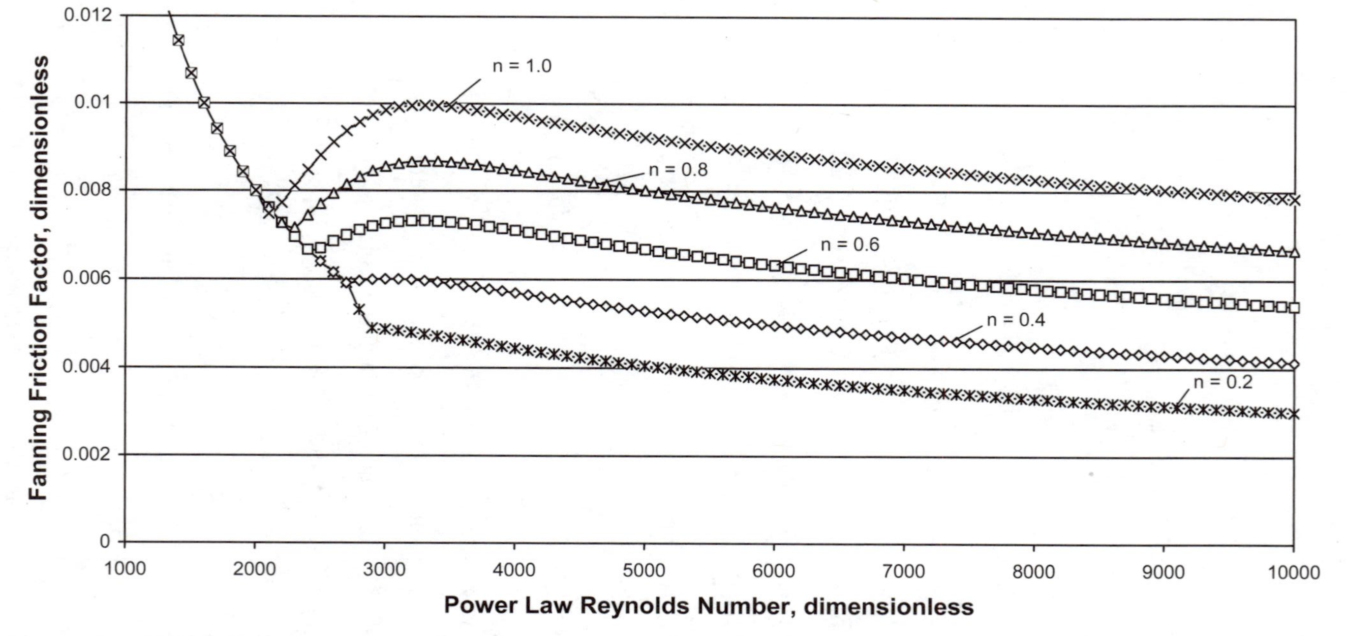

Figure 4.1. Fanning friction factors at different values of the flow behavior index based on Eq. (4.16) to Eq. (4.21) for power law Reynolds numbers to 10,000. [n = 1.0 for Newtonian fluids]

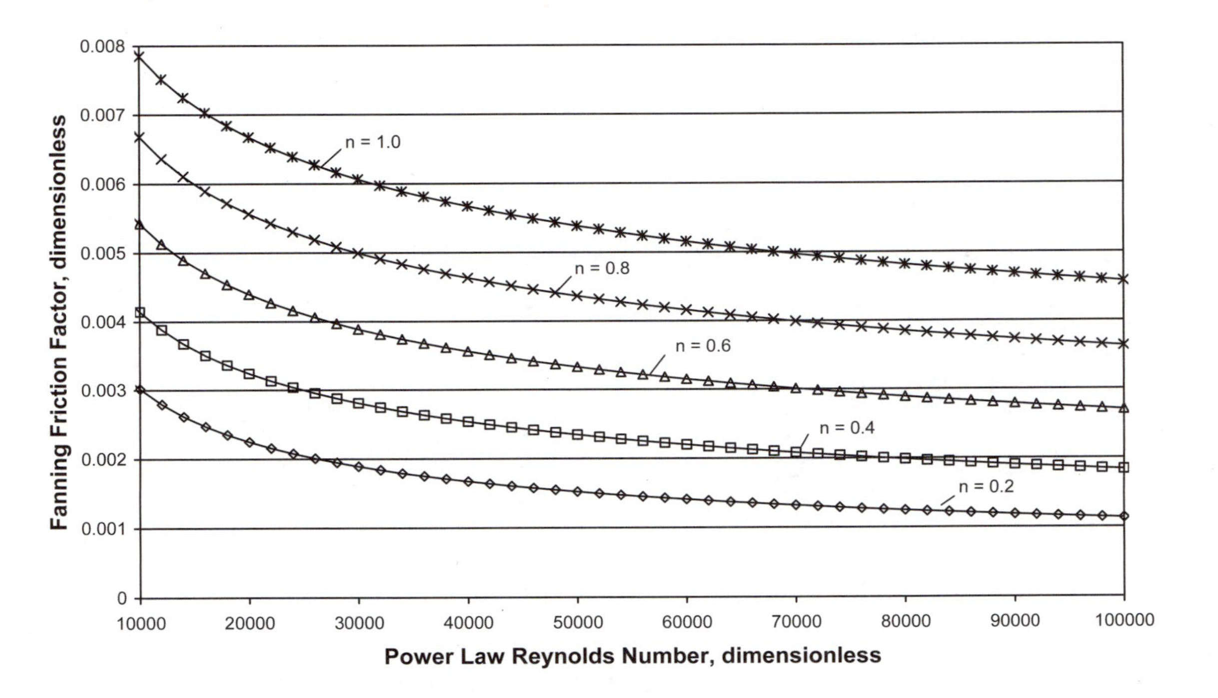

Figure 4.2. Fanning friction factors at different values of the flow behavior index based on Eq. (4.16) to Eq. (4.21) for power law Reynolds numbers from 10,000 to 100,000. [n = 1.0 for Newtonian fluids]

The velocity equations given in the following sections are for fully developed, undisturbed flow in straight, horizontal tubes. Real processing systems contain many elements (valves, tees, elbows, etc.) that cause fluid mixing during flow. In addition, pipe vibration caused by energy inputs from pumps and homogenizers may contribute to mixing. Given these facts, the equations offered below can only be used as general guidelines in examining velocity profiles and estimating the maximum velocity present during tube flow.

Newtonian Fluids. For a Newtonian fluid in laminar flow ($N_{Re} < 2100$), the velocity as a function of the radius is

$$u = \frac{\Delta P}{4L\mu}\left(R^2 - r^2\right) = 2\bar{u}\left(1 - \frac{r^2}{R^2}\right) \tag{4.22}$$

The maximum velocity, found at the center of the pipe, is twice the average velocity:

$$u_{\max} = 2\bar{u} \tag{4.23}$$

Velocity in the turbulent core ($N_{Re} > 4100$) is described with the universal velocity profile [Brodkey, R.S. and H.C. Hershey. 1988. *Transport Phenomena.* McGraw-Hill Book Company, New York]:

$$u^+ = 5.5 + 5.756 \log_{10}\left(y^+\right) \qquad \textit{for } y^+ \geq 30 \tag{4.24}$$

where:

$$u^+ = \frac{u}{u^*} \tag{4.25}$$

$$y^+ = \frac{y u^* \rho}{\mu} \tag{4.26}$$

$$u^{*}=\left(\frac{\sigma_{w}}{\rho}\right)^{1/2}=\bar{u}\left(\frac{f}{2}\right)^{1/2} \tag{4.27}$$

$$y=R-r \tag{4.28}$$

u is the velocity at r, and f is the Fanning friction factor. A simple correlation equation giving the maximum velocity in turbulent flow [Edgerton and Jones, 1970. J. Dairy Science 53: 1353-1357] is also available:

$$u_{\max}=\frac{\bar{u}}{0.0336\log_{10}\left(N_{\mathrm{Re}}\right)+0.662} \tag{4.29}$$

Eq. (4.29) is used in Example Problem 8.11 to determine the maximum fluid velocity in a hold tube so the residence time and thermal history of the fastest moving particle can be established during fluid pasteurization.

Power Law Fluids. For a power law fluid in laminar flow [$N_{\mathrm{Re},PL}<2100+875(1-n)$], the velocity is a function of the radius (r):

$$u=\left(\frac{\Delta P}{2KL}\right)^{1/n}\left(\frac{n}{n+1}\right)\left(R^{\frac{n+1}{n}}-r^{\frac{n+1}{n}}\right) \tag{4.30}$$

or, in terms of the average velocity, as

$$\frac{u}{\bar{u}}=\left(\frac{3n+1}{n+1}\right)\left(1-\left(\frac{r}{R}\right)^{\frac{n+1}{n}}\right) \tag{4.31}$$

The maximum velocity, found at the center of the pipe where r = 0, depends on the value of the flow behavior index:

$$u_{\max}=\bar{u}\left(\frac{3n+1}{n+1}\right) \tag{4.32}$$

Velocity in the turbulent core is found [Skelland, A.H.P. 1967. *Non-Newtonian Flow and Heat Transfer*, John Wiley and Sons, NY] using the universal velocity profile:

$$u^{+}=\left(\frac{5.66}{n^{0.75}}\right)\log\left(y^{+}\right)-\frac{0.566}{n^{1.2}}+\frac{3.475}{n^{0.75}}\left[1.960+0.815n-1.628n\log_{10}\left(3+\frac{1}{n}\right)\right] \tag{4.33}$$

where:

$$u^{+}=\frac{u}{u^{*}} \tag{4.34}$$

$$y^{+}=\frac{y^{n}\left(u^{*}\right)^{2-n}\rho}{K} \tag{4.35}$$

and

$$u^{*}=\left(\frac{\sigma_{w}}{\rho}\right)^{1/2}=\bar{u}\left(\frac{f}{2}\right)^{1/2} \tag{4.36}$$

$$y=R-r \tag{4.37}$$

u in Eq (4.34) is the velocity at r, and f is the Fanning friction factor. A simplified expression for the universal velocity profile of shear-thinning fluids was published in [Clapp, R.M. 1961. International Developments in Heat Transfer, ASME, Part III, Sec. A., pg. 652-661]:

$$u^{+}=\frac{2.78}{n}2.303\log_{10}\left(y^{+}\right)+\frac{3.80}{n} \tag{4.38}$$

for $0.698 < n < 0.813$ and $5480 < N_{Re,Pl} < 42{,}800$

5 Friction Loss Coefficients

5.1 Losses in Standard Valves and Fittings

In discussing energy loss calculations in pipelines, it was noted that friction losses in valves and fittings could be calculated using the velocity head method based on the friction loss coefficient (k_f):

$$F = \frac{k_f \bar{u}^2}{2} \tag{5.1}$$

or the equivalent length (L_e) method:

$$F = \frac{2 f \bar{u}^2 L_e}{D} \tag{5.2}$$

Setting Eq. (5.1) equal to Eq. (5.2), and solving for the equivalent length yields

$$L_e = \frac{k_f D}{4 f} \tag{5.3}$$

Equivalent lengths are usually calculated from the numerical value of friction loss coefficient, determined from experimental data. Since the Fanning friction factor is a function of the Reynolds number, the L_e value is also a function of the Reynolds number. Given this, and the fact that L_e values are generally not given as a function of the Reynolds number, employing k_f values and Eq. (5.1) is recommended for friction loss calculations. Friction loss coefficients are calculated in Example Problems 8.7 and 8.8.

Friction loss coefficients for standard valves and fittings may be estimated using the following equation [Hooper, W.B. 1981 (August 24). Chemical Engineering, 96-100]:

$$k_f = \frac{k_1}{N_{\text{Re}}} + k_\infty \left(1 + \frac{1}{D_{inch}}\right) \tag{5.4}$$

where: D_{inch} = internal diameter of the attached pipe in units of inches. Values of k_1 and k_∞ were determined from experimental data and are summarized in Table 5.1. $N_{Re,PL}$ should be used in place of N_{Re} for power law fluids. Eq. (5.4) is acceptable for the pipe diameters normally found in food and pharmaceutical pipelines. If dealing with very large diameter pipelines (greater than 6 inches), the method published by R. Darby [2001, March. Chemical Engineering, 66-73] should be considered. The equation and constants for this procedure, known as the 3-k method, are given in Appendix 9.12.

Limited data have been published describing the pressure loss in valves and fittings during the laminar flow of non-Newtonian, biological fluids. Using applesauce at various dilutions, Steffe et al. [1984, Trans. ASAE 27: 616-619] gave three equations (Table 5.2) for sanitary components for $20 < N_{\text{Re},PL} < 700$ in 1.5 inch stainless steel tubing (D_i = 1.50 inch). Calculating the friction loss coefficients using the comparable method suggested by Hooper [Eq. (5.4)] produced very conservative k_f values: numbers were 2 to 3 times larger than those found using the equations in Table 5.2.

Friction loss coefficients from tank exits and entrances may be predicted as

$$k_f = \frac{k_1}{N_{\text{Re}}} + k_\infty \tag{5.5}$$

Values of k_1 and k_∞ are summarized in Table 5.3 and 5.4 for different exit scenarios. Friction loss coefficients for tapered and square pipe expansions and contractions may be found in Tables 5.5 to 5.8, respectively; and dimensions of typical reducers or expanders used in sanitary tubing are given in Table 5.9.

5.2 Losses in Equipment Based on Data for Water

Companies often provide experimental data describing the pressure loss occurring with fluid flow through their equipment. These data are usually collected for water, and plotted as pressure loss (psi or inches of water) versus volumetric flow rate (gpm). Considering the friction factors for each fluid, one can make an approximation of the friction loss that occurs with other Newtonian fluids (more or less viscous than water), as well as non-Newtonian fluids.

Assume that the energy loss per unit mass ($\Delta P/\rho$) can be expressed in terms of the Fanning friction factor (f) and the equivalent length of the equipment (L_e). Then, the equation for the application fluid is

$$\left(\frac{\Delta P}{\rho}\right)_{application\ fluid} = \left(\frac{2 f \bar{u}^2 L_e}{D}\right)_{application\ fluid} \tag{5.6}$$

and, the equation for water is

$$\left(\frac{\Delta P}{\rho}\right)_{water} = \left(\frac{2 f \bar{u}^2 L_e}{D}\right)_{water} \tag{5.7}$$

Assuming the equivalent length is the same in each case (the other factors are clearly the same); and dividing Eq. (5.6) by Eq. (5.7) yields:

$$\frac{\left(\frac{\Delta P}{\rho}\right)_{application\ fluid}}{\left(\frac{\Delta P}{\rho}\right)_{water}} = \frac{f_{application\ fluid}}{f_{water}} \tag{5.8}$$

where the friction factors are those for the connecting pipes. Hence, losses for a non-Newtonian fluid can be estimated from experimental data for water by including the friction factor ratio of the two fluids:

$$\left(\frac{\Delta P}{\rho}\right)_{application\ fluid} = \left(\frac{\Delta P}{\rho}\right)_{water}\left(\frac{f_{application\ fluid}}{f_{water}}\right) \tag{5.9}$$

For the special case of Newtonian fluids in laminar flow, this becomes:

$$\left(\frac{\Delta P}{\rho}\right)_{application\ fluid} = \left(\frac{\Delta P}{\rho}\right)_{water}\left(\frac{\mu_{application\ fluid}}{\mu_{water}}\right) \tag{5.10}$$

Eq. (5.9) is applied to a pneumatic valve and a strainer, for both a Newtonian fluid and a non-Newtonian fluid, in Example Problems 8.7 and 8.8, respectively.

Table 5.1. Values of k_1 and k_∞ for standard valves and fittings.

Valve or Fitting [#]	k_1	k_∞
elbow 90°: standard (*R/D* = 1), threaded	800	0.40
elbow 90°: standard (*R/D* = 1), flanged or welded	800	0.25
elbow 90°: long radius (*R/D* = 1.5), all types	800	0.20
elbow 45°: standard (*R/D* = 1), all types	500	0.20
elbow 45°: long radius (*R/D* = 1.5), all types	500	0.15
elbow 180°: standard (*R/D* = 1), threaded	1000	0.60
elbow 180°: standard (*R/D* = 1), flanged or welded	1000	0.35
elbow 180°: long radius (*R/D* = 1.5), all types	1000	0.30
tee used as elbow: standard, threaded	500	0.70
tee used as elbow: long radius, threaded	800	0.40
tee used as elbow: standard, flanged or welded	800	0.80
tee (run through): threaded	200	0.10
tee (run through): flanged or welded	150	0.50
gate valve (open)	300	0.10
globe valve, standard (open)	1500	4.00
globe valve, angle or Y-type (open)	1000	2.00
diaphragm valve, dam-type (open)	1000	2.00
butterfly	800	0.25
check valve (lift type)	2000	10.00
check valve (swing type)	1500	1.50
check valve (tilting-disk)	1000	0.50

[#] Except for some high-pressure applications, threaded product contact surfaces are not allowed in sanitary systems (3A Standard Number 63-10). Standard threaded steel is acceptable for water supply systems.

Table 5.2. Friction loss coefficients for applesauce in laminar flow.

Valve or Fitting	Friction Loss Coefficient (-)
plug valve, branch flow	$k_f = 30.3\left(N_{\text{Re},PL}\right)^{-0.5}$
tee through branch (as elbow)	$k_f = 29.4\left(N_{\text{Re},PL}\right)^{-0.5}$
elbow 90°, flanged	$k_f = 191\left(N_{\text{Re},PL}\right)^{-0.9}$

Table 5.3. k_1 and k_∞ for square entrance or exit from tank to pipe.

	k_1	k_∞
pipe entrance: flush, square *tank* *pipe* flow 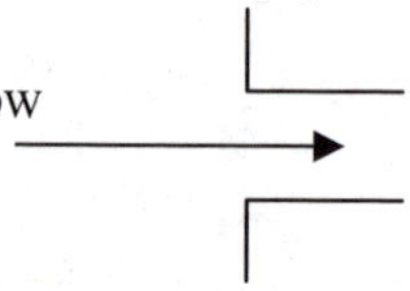	160	0.5
pipe entrance: inward-projecting (Borda) *tank* *pipe* flow 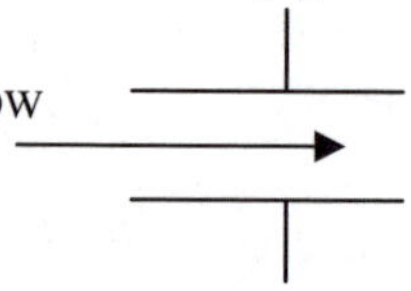	160	1.0
pipe exit (all geometries)	0	1.0

Source: Source: Hooper, W.B. 1981. Chemical Engr. Aug. 24, 96-99.

Table 5.4. Values of k_1 and k_∞ for rounded exit from tank to pipe.

	r/D	k_1	k_∞
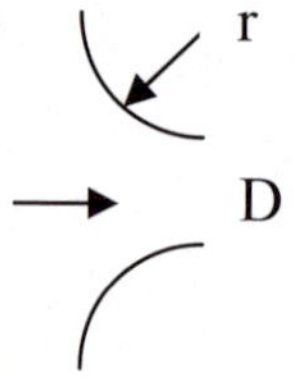 r D	0.0 (sharp)	160 for all *r/D*	0.5
	0.02		0.28
	0.04		0.24
	0.06		0.15
	0.10		0.09
	0.15 (and up)		0.04

Source: Darby, R. 2001. *Chemical Engineering Fluid Mechanics (Second edition*). Marcel Dekker, New York. pg. 213.

Table 5.5. Friction loss coefficients for tapered pipe expansion.

D_1, D_2, θ	Tapered Pipe Expansion
Inlet N_{Re}	k_f based on inlet velocity
$0° < \theta < 45°$ and $N_{\text{Re}} \leq 4000$	$k_f = 2\left[1-\left(\frac{D_1}{D_2}\right)^4\right]\left[2.6\sin\left(\frac{\theta}{2}\right)\right]$
or	or
$N_{\text{Re}} > 4000$	$k_f = [1+3.2f]\left\{\left[1-\left(\frac{D_1}{D_2}\right)^2\right]^2\right\}\left[2.6\sin\left(\frac{\theta}{2}\right)\right]$
$45° < \theta \leq 180°$ and $N_{\text{Re}} \leq 4000$	$k_f = 2\left[1-\left(\frac{D_1}{D_2}\right)^4\right]$
or	or
$N_{\text{Re}} > 4000$	$k_f = [1+3.2f]\left\{\left[1-\left(\frac{D_1}{D_2}\right)^2\right]^2\right\}$

Source: Hooper, W.B. 1988. Chemical Engr. Nov. 7, 89-92.

N_{Re} is the upstream Reynolds number; f = Fanning friction factor calculated at the upstream Reynolds number; use $N_{Re,PL}$ in place of N_{Re} for power law fluids; θ = angle, degrees. Note: θ for sanitary expansions fall in the 0 to 45 degree range.

Table 5.6. Friction loss coefficients for square pipe expansion.

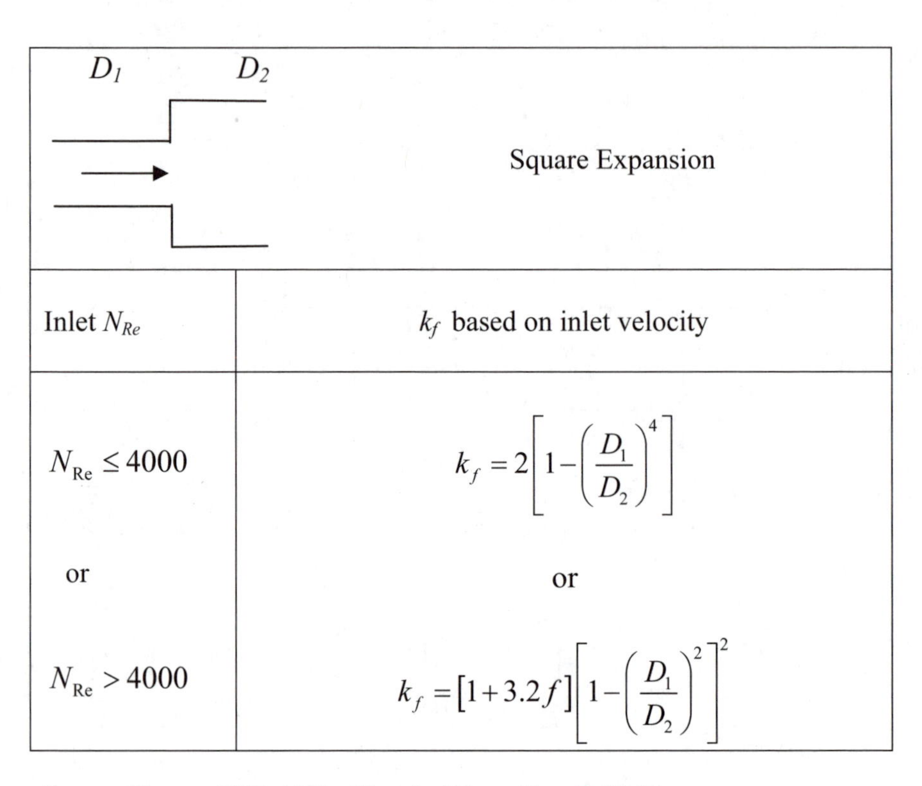

Square Expansion	
Inlet N_{Re}	k_f based on inlet velocity
$N_{\text{Re}} \leq 4000$	$k_f = 2\left[1 - \left(\frac{D_1}{D_2}\right)^4\right]$
or	or
$N_{\text{Re}} > 4000$	$k_f = \left[1 + 3.2f\right]\left[1 - \left(\frac{D_1}{D_2}\right)^2\right]^2$

Source: Hooper, W.B. 1988. Chemical Engr. Nov. 7, 89-92.

N_{Re} is the upstream Reynolds number; f = Fanning friction factor calculated at the upstream Reynolds number; use $N_{Re,PL}$ in place of N_{Re} for power law fluids.

Table 5.7. Friction loss coefficients for tapered pipe contractions.

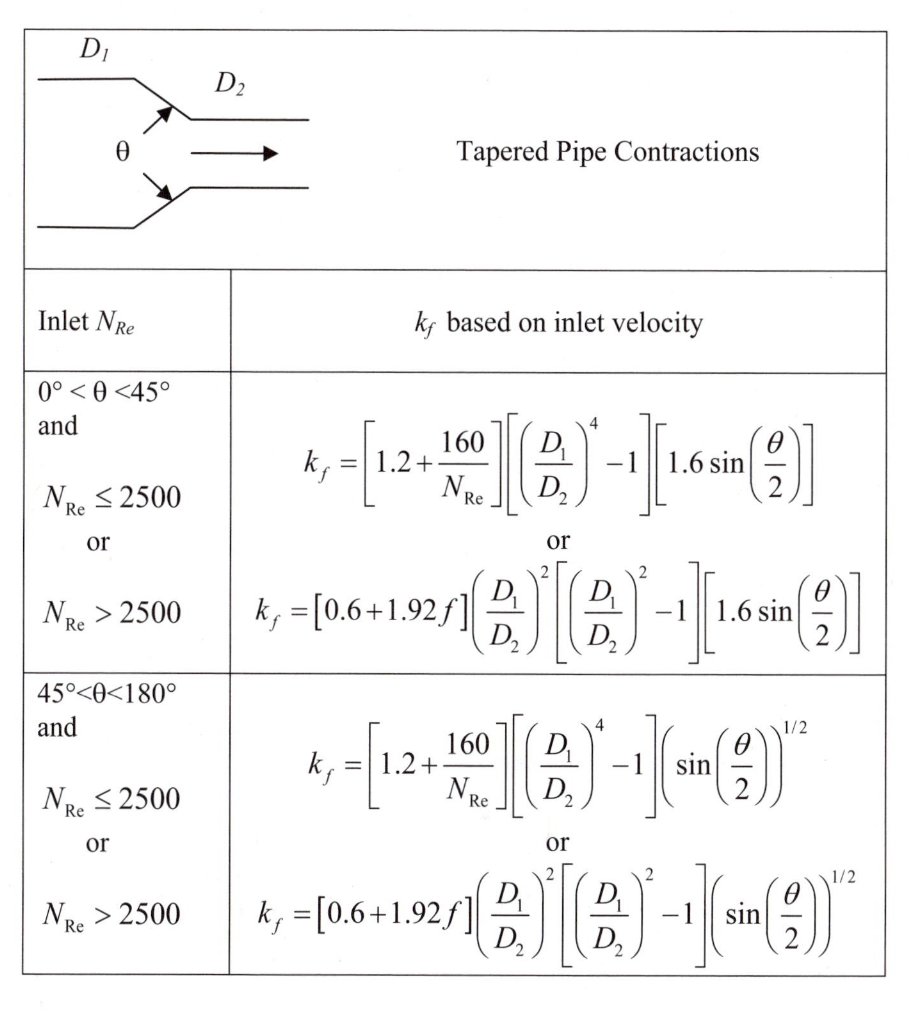

D_1 D_2 θ	Tapered Pipe Contractions
Inlet N_{Re}	k_f based on inlet velocity
$0° < \theta < 45°$ and $N_{\mathrm{Re}} \le 2500$	$k_f = \left[1.2 + \frac{160}{N_{\mathrm{Re}}}\right]\left[\left(\frac{D_1}{D_2}\right)^4 - 1\right]\left[1.6\sin\left(\frac{\theta}{2}\right)\right]$
or	or
$N_{\mathrm{Re}} > 2500$	$k_f = \left[0.6 + 1.92f\right]\left(\frac{D_1}{D_2}\right)^2\left[\left(\frac{D_1}{D_2}\right)^2 - 1\right]\left[1.6\sin\left(\frac{\theta}{2}\right)\right]$
$45° < \theta < 180°$ and $N_{\mathrm{Re}} \le 2500$	$k_f = \left[1.2 + \frac{160}{N_{\mathrm{Re}}}\right]\left[\left(\frac{D_1}{D_2}\right)^4 - 1\right]\left(\sin\left(\frac{\theta}{2}\right)\right)^{1/2}$
or	or
$N_{\mathrm{Re}} > 2500$	$k_f = \left[0.6 + 1.92f\right]\left(\frac{D_1}{D_2}\right)^2\left[\left(\frac{D_1}{D_2}\right)^2 - 1\right]\left(\sin\left(\frac{\theta}{2}\right)\right)^{1/2}$

Source: Hooper, W.B. 1988. Chemical Engr. Nov. 7, 89-92.

N_{Re} is the upstream Reynolds number; f = Fanning friction factor calculated at the upstream Reynolds number; use $N_{Re,PL}$ in place of N_{Re} for power law fluids; θ = angle, degrees. Note: θ for sanitary expansions fall in the 0 to 45 degree range.

Table 5.8. Friction loss coefficients for square pipe contraction.

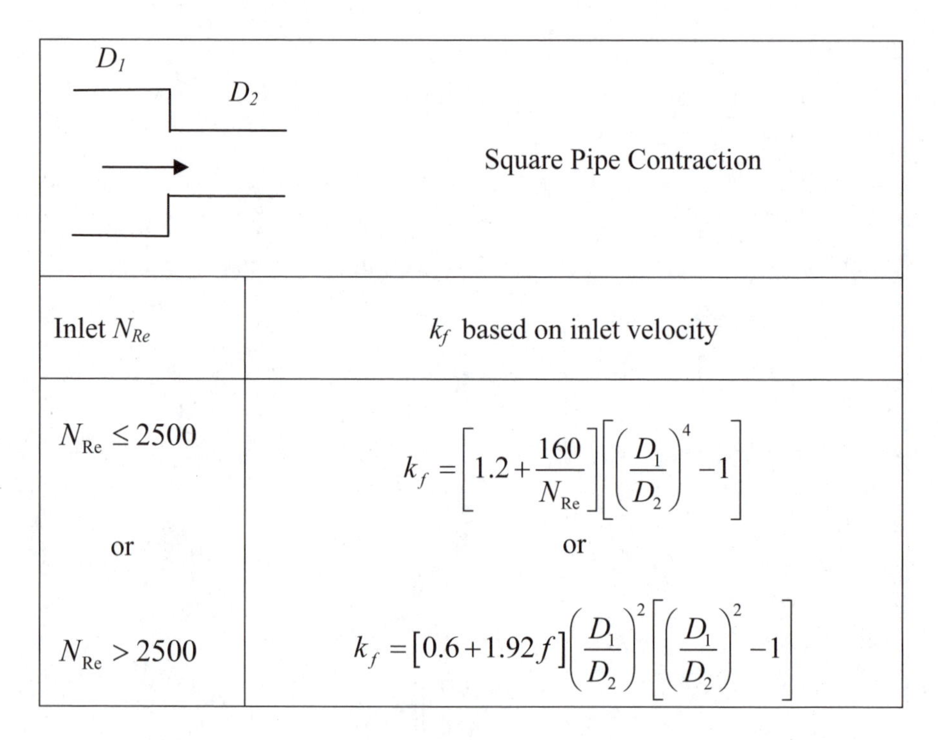

D_1 → D_2	Square Pipe Contraction
Inlet N_{Re}	k_f based on inlet velocity
$N_{\mathrm{Re}} \leq 2500$	$k_f = \left[1.2 + \dfrac{160}{N_{\mathrm{Re}}}\right]\left[\left(\dfrac{D_1}{D_2}\right)^4 - 1\right]$
or	or
$N_{\mathrm{Re}} > 2500$	$k_f = \left[0.6 + 1.92 f\right]\left(\dfrac{D_1}{D_2}\right)^2\left[\left(\dfrac{D_1}{D_2}\right)^2 - 1\right]$

Source: Hooper, W.B. 1988. Chemical Engr. Nov. 7, 89-92.

N_{Re} is the upstream Reynolds number; f = Fanning friction factor calculated at the upstream Reynolds number; use $N_{Re,PL}$ in place of N_{Re} for power law fluids.

Table 5.9. Dimensions of typical reducers or expanders for sanitary tubing.

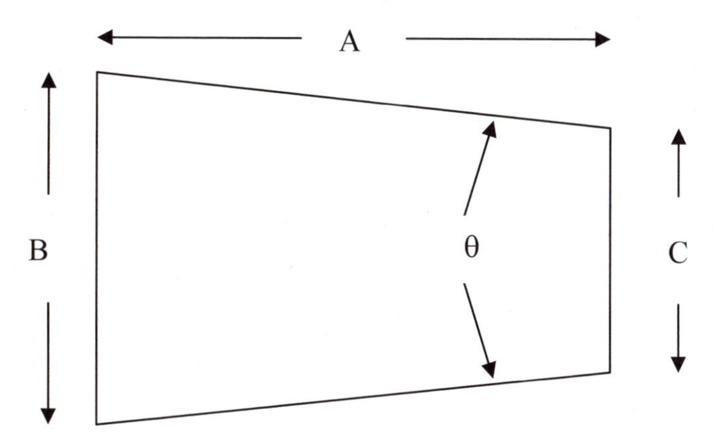

Size	A, inches	B, inches	C, inches	θ, degrees
¾" x ½"	2	0.75	0.5	7.1
1" x ½"	2.5	1	0.5	11.4
1" x ¾"	2	1	0.75	7.1
1½" x ½"	3.5	1.5	0.5	16.2
1½" x ¾"	4	1.5	0.75	10.7
1½" x 1"	3	1.5	1	9.5
2" x ¾"	4.25	2	0.75	16.7
2" x 1"	5	2	1	11.4
2" x 1½"	3	2	1.5	9.5
2½" x 1½"	5	2.5	1.5	11.4
2½" x 2"	3	2.5	2	9.5
3" x 1½"	7	3	1.5	12.2
3" x 2"	5	3	2	11.4
3" x 2½"	3	3	2.5	9.5
4" x 2"	9.125	4	2	12.5
4" x 2½"	7.125	4	2.5	12.0
4" x 3"	5.125	4	3	11.1

Note $\theta = 180°$ for a sharp edged entrance or exit.

6 Handling Shear-Sensitive Fluids

6.1 Shear-Sensitive Fluids

Many biological fluids are shear-sensitive (fragile) and easily damaged during processing. Typical problems include the breakdown of emulsions, the disruption of microbial colonies in fermentation systems, and damage to particulates during handling. Shear-sensitivity of a fluid may also contribute to beneficial processing objectives by creating emulsions, or enhanced aggregation of colloids or microorganisms. In either case, the techniques presented in this chapter enable one to quantitatively characterize the mechanical components of the process. These methods are valuable tools for scaling-up (or down) processing systems, for considering alternative process designs, and for controlling quality factors resulting from the shear-sensitivity of processed fluids.

Viscous dissipation of energy is found whenever fluids are in motion, and every component of a pipeline processing system causes fluid agitation or mixing. This mixing is intentional when stirred tanks or static (in-line pipe mixers) are included in the process. Unintentional, and often intense, shearing may also occur in fittings, valves, heat exchangers, strainers, and other typical pipeline components. Pumps may also contribute significantly to mixing. Centrifugal pumps, which employ rapidly rotating impellers to accelerate the fluid within the pump casing, generate strong fluid agitation and concomitant damage to shear-sensitive materials.

Methods to estimate shear work, shear power intensity, and average shear rate for particulate transport in a handling system are presented in this chapter. These techniques are needed to account for the shear inputs of the mechanical components of a processing system, and investigate how shear-sensitivity of a material influences product manufacturing and quality.

6.2 Shear Work (W_s)

Shear work (W_s) deals with energy changes in a mechanical system that expose a fluid to a shear environment. The term "shear work" is used to distinguish it from work inputs, such as those contributing to changes in potential energy, which do not shear the product. Contributions from all fluid handling elements must be considered to evaluate shear work in a complete processing system.

Shear work introduced from a mixing tank depends on whether or not the tank is operated on a batch or a continuous (constant throughput) basis: 1) In the batch mode, W_s is the product of the power input from the mixer (Φ) multiplied by the mixing time (t) divided by the mass (m) of fluid in the tank; 2) In a continuous system, W_s is simply the power input divided by the mass flow rate ($\dot{m}$) of fluid through the tank (Table 6.1). Calculations for typical pipeline components involve the familiar terms found in the mechanical energy balance equation (Eq. 3.2), and are summarized in Table 6.1. These values can be predicted or, if the pressure drop over any particular element of the system (including valves, fittings and straight pipe) is known, W_s can be calculated as $\Delta P/\rho$. Considering all mechanical elements of a process in contact with the fluid, the total shear work for a system is

$$W_{s,\text{total}} = \sum \frac{\Phi}{\dot{m}} + \sum \frac{k_f \bar{u}^2}{2} + \sum \frac{2 f \bar{u}^2 L}{D} + \sum \left(\frac{\Delta P}{\rho} \right) \qquad (6.1)$$

Eq. (6.1) includes the possibility of multiple (continuous flow) tanks, numerous valves and fittings, transport pipes with different diameters, and multiple pieces of equipment including the pump.

Table 6.1. Shear work equations for different components of a fluid pumping system.

System Component	Shear Work (W_s)
Tank: batch operation	$W_s = \frac{\Phi t}{m}$
Tank: continuous flow	$W_s = \frac{\Phi}{\dot{m}}$
Standard valves or fittings	$W_s = \frac{k_f \overline{u}^2}{2}$
Straight pipe	$W_s = \frac{2 f \overline{u}^2 L}{D}$
Equipment: pneumatic valves, heat exchangers, strainers, centrifugal pumps, etc.	$W_s = \frac{\Delta P}{\rho}$

W_s has units of J/kg, and represents the energy change per unit mass of the product as it travels through a system component. Pressure increase or decrease through a mechanical component is directly proportional to shear work. Pressure decreases as the fluid loses energy, such as when it travels through a valve. Conversely, the pressure increases as the fluid gains energy when passing through a pump. In each case, significant mechanical agitation occurs; hence, the pressure differential should always be considered a positive number when evaluating shear work input because it represents a change in energy that contributes to product shearing. Placing pressure transducers before and after critical equipment provides a measure of the pressure differential (ΔP) across the unit. If not determined from known pressure differences, shear work must be predicted (Table 6.1) using the Fanning friction factor (f) or a friction loss coefficient (k_f).

In examining a process, both the shear work for each individual component and the total shear work for the system should be calculated. It is particularly important to conduct a careful examination of the pump before deciding to include it as a contributing factor. Centrifugal pumps generate very large friction losses because they operate by converting kinetic energy to pressure energy; however, low shear pumps (such as rotary lobe and gear pumps), and very low shear pumps (such as diaphragm and progressing cavity pumps), can generate very large differential pressures by fluid compression but may add little shear work to the product. Only pumps producing significant friction losses in the fluid should be included in the calculation of total shear work.

6.3 Shear Power Intensity (S)

In addition to shear work, the shear power intensity (S) for system components must be considered. This parameter is defined for a particular component as the product of the shear work times the mass flow rate of the product divided by the fluid volume of the component:

$$S = \frac{W_s \, \dot{m}}{V} \tag{6.2}$$

where V is the volumetric capacity of the component under static conditions. The shear power intensity for a tank is equal to the power divided by the volume of fluid in the tank (Φ/V) for both batch and continuous operations.

Shear power intensity has units of J $s^{-1}m^{-3}$ or W m^{-3}, and represents the change in power per unit volume (rate of energy change per unit volume) given to the fluid as it travels through a particular component. It provides a measure of the rate of energy change per unit volume in different mechanical parts of the system, and helps distinguish sections of the system that may contribute high shear work inputs but low shear intensities. This is an important distinction because areas can have high shear work inputs (long sections of straight pipe for example) but may contribute little to product degradation if the shear power intensity is low. Example Problem 8.9 illustrates this concept. Shear power intensity generated by a centrifugal pump is calculated in Example Problem 8.10.

The static volume, V in Eq. (6.2), is clearly defined for most equipment: It is simply the void volume of the physical object normally occupied by fluid. There is no obvious static volume to use in calculating S for the convergence at a pipe entrance, divergence at a pipe exit, or for flow through an orifice. Although mixing length theory (coupled with experimental research) could be employed to develop a rational definition of V, it requires various assumptions that make it impractical for engineering practice. A thorough investigation of the pipe exit is even more complicated. To facilitate the analysis, the volume term for the pipe entrance, the pipe exit and the orifice is assumed to be the space occupied by a length of pipe equal to two pipe diameters in length: $V = \pi D^3/2$. For a convergence or divergence involving two pipes, use the diameter of the smaller pipe when calculating V.

6.4 Critical Values of W_s and S

The combination of the ingredients and the processing equipment used to manufacture any product is unique. Critical values of shear work (W_s) and shear power intensity (S) are also unique to each product and manufacturing system. Critical values may be the maximum or minimum values required for a successful operation. To further clarify and expand the discussion in the previous sections, consider the unloading and storage of a shear-sensitive fluid, illustrated in Fig. 6.1. The unloading operation must account for the following components: pipe entrance as fluid leaves the truck, pump, valve, strainer, fluid exit from the pipe as it enters the tank, three elbows, and the total length of pipe required for fluid transport; and storage conditions must account for agitation in the tank.

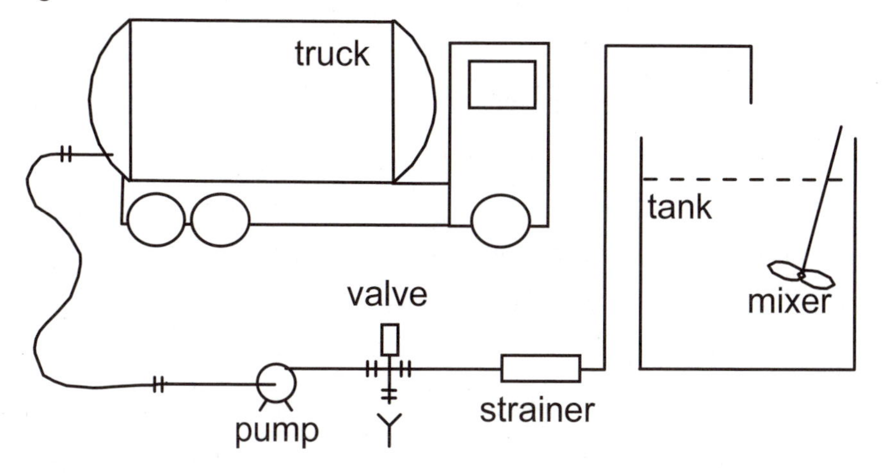

Figure 6.1. Mechanical system for a shear-sensitive fluid.

The components of the system (Fig. 6.1), and the variables that must be considered when evaluating shear-sensitivity, are summarized in Table 6.2. Pressure drop over the pump is determined from direct measurement, or by applying the mechanical energy balance equation to the entire system. Fig. 6.1 is similar to Fig. 3.3 and the results for the work input (W) required by the pump (Table 3.1) would allow the pressure increase over the pump to be calculated: $\Delta P = W\rho$. This calculation is included in Example Problem 8.7. The pressure drop information for the strainer would be obtained from the manufacturer – these data are based on water transport but can be corrected for other fluids using Eq. (5.9) or Eq. (5.10). W_s for the valve could also be calculated using pressure loss data for water (common for pneumatic valves) or with Eq. (5.4) using the appropriate friction loss coefficient.

Friction loss coefficients are needed to find the shear work and shear power intensity for three elbows, the tank entrance, and the tank exit. Calculation of W_s and S for the pipe requires the Fanning friction factor and L, the total length of pipe in the system. The volumetric fill capacity (V) of a component may be given or measured as gallons (or liters) of water. This information is converted into spatial volume using the density of water.

Fluid is pumped from the truck into a storage tank equipped with a marine style impellor for mixing (Fig. 6.1). The storage tank should be treated as an independent system. Shear work input to the fluid that occurs during mixing depends on the power input (provided via the mixer motor) and the total mixing time (t). Shear power intensity is simply the power input divided by the volume of fluid in the tank (Table 6.2).

Table 6.2. Shear work (W_s), shear power intensity (S) and volume (V, required in the calculation of shear power intensity) for the system illustrated in Fig. 6.1.

	W_s	S	V
Unloading			
pipe entrance	$\frac{k_{f,entrance}\overline{u}^2}{2}$	$\frac{W_s\dot{m}}{V}$	$\frac{\pi D^3}{2}$
pump	$\frac{\Delta P}{\rho}$	$\frac{W_s\dot{m}}{V}$	pump fill volume
valve	$\frac{k_{f,valve}\overline{u}^2}{2}$	$\frac{W_s\dot{m}}{V}$	valve fill volume
strainer	$\frac{\Delta P}{\rho}$	$\frac{W_s\dot{m}}{V}$	strainer fill volume
elbows (3)	$\frac{3k_{f,elbow}\overline{u}^2}{2}$	$\frac{3W_s\dot{m}}{V}$	3 x elbow fill volume
pipe exit	$\frac{k_{f,exit}\overline{u}^2}{2}$	$\frac{W_s\dot{m}}{V}$	$\frac{\pi D^3}{2}$
straight pipe	$\frac{2f\overline{u}^2L}{D}$	$\frac{W_s\dot{m}}{V}$	$\frac{\pi D^2L}{4}$
Storage			
storage tank	$\frac{\Phi t}{m}$	$\frac{\Phi}{V}$	fluid volume in tank

W_s should be calculated for individual components of the system (Table 6.2) and, because the work effect may be cumulative, it should also be calculated for the total system:

$$W_{s,\mathrm{total}} = \left(\frac{\Delta P}{\rho}\right)_{pump} + \frac{k_{f,valve}\,\overline{u}^2}{2} + \left(\frac{\Delta P}{\rho}\right)_{strainer} + \frac{3k_{f,elbow}\,\overline{u}^2}{2} + \frac{k_{f,exit}\,\overline{u}^2}{2} + \frac{2f\,\overline{u}^2 L}{D} + \frac{\Phi t}{m} \tag{6.3}$$

Although it is possible to calculate an overall S value ($W_{s,\mathrm{total}} / V_{\mathrm{total}}$), this parameter has limited value for characterizing a cumulative effect.

Since every product is different, objective measures of product quality must be correlated to the critical values of shear work input and shear power intensity. The effect of fluid transport on an emulsion, such as mayonnaise, could be studied by removing samples at different points in the process and checking for emulsion stability, or monitoring particle size distribution. Excessive energy input may, for example, cause oil droplets to coalesce resulting in the breakdown of the emulsion. Breakdown criteria could be established by rheological, microscopic, or colorimetric analyses. These quality criteria could be correlated to the shear work and shear power intensity found in different components of the processing system to determine critical values or, in this example, maximum permissible levels. The mechanical system could then be designed to stay within those limits. A similar approach could be used to investigate shear induced aggregation in colloidal solutions, disruption of microbial colonies in fermentation broths, or structural breakdown in yoghurt.

There are a number of factors that affect W_s and S during a process. Flow rate is obviously the most important factor since work and power input levels will increase at higher levels of flow. Fluid temperature during processing is also very important since rheological properties are a strong function of temperature, and this may alter the influence of different work inputs and shear power levels on fluid quality. As stated previously, each product and process is unique; therefore, the individual characteristics of any particular system must be carefully examined.

6.5 Shear-Sensitive Particulates

In addition to shear work input and shear power intensity, it may be necessary to consider the average shear rates that particles experience in passing through system components. Large particulates, such as diced vegetables and pasta pieces, may be damaged if forced through small spaces at high speeds. The average shear rate through a system component can be used to benchmark this problem. It can be estimated as two times the mean velocity divided by the characteristic dimension (*d*) of the component:

$$\dot{\gamma}_{avg} = \frac{2\overline{u}}{d} \tag{6.4}$$

where the characteristic dimension is the minimum clearance length in the cross-sectional flow area. In the simple case of a pipe, the characteristic dimension is the internal diameter. A reasonable value for a centrifugal pump would be the distance between the tip of the impeller and the inner casing of the pump. High average shear rates in a fluid handling system may damage large particulates.

Average shear rate defined by Eq. (6.4) is given as a tool to examine potential particulate damage during transport. It has a very different meaning than the average shear rate (defined by Eq. 2.12) used in mixer viscometry. In a large mixing tank, maximum shear rate would be a better predictor of particulate damage. This could, for example, be calculated in a tank with an anchor impeller as the tip speed of the impeller divided by the minimum distance between the tip and the tank.

6.6 Scale-Up Calculations

Scaling Pipelines. Pipeline scale-up is a common problem in bioprocess engineering. In this analysis, assume a pipe is used to transport a shear-sensitive fluid, and an examination of the process showed that one set of flow parameters (case 1) produces an acceptable product. Furthermore, assume a new process (case 2) was requested that increases the mass flow rate by a factor of *C,* defined as the ratio of the two mass flow rates:

$$C = \frac{\dot{m}_{\text{case 2}}}{\dot{m}_{\text{case 1}}} \tag{6.5}$$

Maintaining a constant shear power intensity, one must keep S constant for both cases; hence, from Eq. (6.2),

$$\left(\frac{W_s \dot{m}}{V}\right)_{\text{case 1}} = \left(\frac{W_s \dot{m}}{V}\right)_{\text{case 2}} \tag{6.6}$$

If losses in the pump are negligible, the shear intensity only includes straight pipe, and Eq. (6.6) becomes

$$\left(\frac{2 f \overline{u}^2 L}{D} \frac{\dot{m}}{V}\right)_{\text{case 1}} = \left(\frac{2 f \overline{u}^2 L}{D} \frac{\dot{m}}{V}\right)_{\text{case 2}} \tag{6.7}$$

Assume the length is constant and the flow is laminar so $f = 16 / N_{\text{Re}} = 16\mu / \left(\rho D \overline{u}\right)$. Also, recall that $\overline{u} = \dot{m} / \left(A\rho\right)$, and recognize that the fluid properties (μ and ρ) are unchanged by flow rate. Using these facts, and algebraic simplification, Eq. (6.7) becomes

$$\left(\frac{\dot{m}^2}{D^5}\right)_{\text{case 1}} = \left(\frac{\dot{m}^2}{D^5}\right)_{\text{case 2}} \tag{6.8}$$

Substituting the flow rate for case 2 ($\dot{m}_{\text{case 2}} = C\,\dot{m}_{\text{case 1}}$) into Eq. (6.8), and solving for the new diameter gives the desired result:

$$D_{\text{case 2}} = C^{2/5}\left(D_{\text{case 1}}\right) \tag{6.9}$$

This indicates that the shear intensity can be maintained if the tube diameter is increased by a factor of $C^{2/5}$. Consider, for example, the case where the goal is to double the mass flow rate (meaning $C = 2$) but maintain the same shear power intensity values:

$$D_{\text{case 2}} = 2^{2/5}\left(D_{\text{case 1}}\right) = 1.32\left(D_{\text{case 1}}\right) \tag{6.10}$$

Eq. (6.10) shows it is possible to double the flow rate, while keeping the rate of energy input constant, if the diameter of the pipeline is increased by a factor of approximately 1.32. In actual processing systems where significant energy losses also occur in fittings and equipment (including valves, strainers, pumps, heat exchangers, etc.), the analysis is more complex but the principle is the same. Scale-up of a shear-sensitive fluid (cream) in a pipeline is examined in Example Problem 8.9.

Scaling Components Based on Maximum Shear Intensity. Maximum shear power intensity in a processing system will occur in a particular system component such as a valve. It may be important to maintain, or not exceed, this value. Assume, for example, $C = \dot{m}_{\text{case 2}} / \dot{m}_{\text{case 1}}$. To maintain constant shear power intensity, one must keep S constant for both cases: $(S)_{\text{case 1}} = (S)_{\text{case 2}}$. Substituting the appropriate shear work term, in this case the pressure drop divided by the density, into Eq. (6.6) yields

$$\left(\frac{\Delta P}{\rho}\frac{\dot{m}}{V}\right)_{\text{case 1}} = \left(\frac{\Delta P}{\rho}\frac{\dot{m}}{V}\right)_{\text{case 2}} \tag{6.11}$$

To maintain this equality, it may be necessary to increase the static volume of the component. This may be achieved in a valve, for example, by using a larger valve body for the higher flow rate process.

Using the Reynolds Number to Scale Pipe Diameters. In the case of scaling pipelines discussed above (a Newtonian fluid in laminar flow in straight pipe), maintaining a constant Reynolds number means

$$\left(N_{\text{Re}}\right)_{\text{case 1}} = \left(N_{\text{Re}}\right)_{\text{case 2}} \tag{6.12}$$

or

$$\left(\frac{\rho D\bar{u}}{\mu}\right)_{\text{case 1}} = \left(\frac{\rho D\bar{u}}{\mu}\right)_{\text{case 2}} \tag{6.13}$$

Substituting $\bar{u} = 4Q/\left(\pi D^2\right)$ into Eq. (6.13), and recognizing that $\dot{m} = Q\rho$, gives

$$\left(\frac{\dot{m}}{D}\right)_{\text{case 1}} = \left(\frac{\dot{m}}{D}\right)_{\text{case 2}} \tag{6.14}$$

This result is clearly different than Eq. (6.8). It shows that using a constant Reynolds number as the scaling criterion to maintain a constant level of shear input results in an overly large increase in pipe diameter: $D_{\text{case 2}} = C\,D_{\text{case 1}}$ instead of $D_{\text{case 2}} = C^{2/5}\,D_{\text{case 1}}$. Thus, one should be cautious if using the Reynolds number as a scaling criterion to account for shear degradation.

7 Thermal Processing of Biological Fluids

Thermal treatments are commonly used to eliminate pathogens from biological fluids such as food and pharmaceutical products; therefore, a qualitative assessment of these systems may play an important role in the analysis of bioprocessing pipelines. This chapter examines how fast microorganisms die (death kinetics) at a given temperature; and how a variable temperature treatment can be equated to an equivalent time at a constant temperature to determine the overall lethality of a process using the general method. The techniques presented provide a simple means of characterizing the destruction of microorganisms during the heat treatment of a biological fluid. These techniques can also be used to characterize heat treatments to inactivate enzymes (pectinase in orange juice for example), and to examine the destruction of nutrients (thiamine in milk) and various changes in food quality such as texture and color.

7.1 Death Kinetics of Microorganisms

It is common practice to describe the death of microorganisms at a constant temperature as an irreversible first order reaction involving the number of organisms (N), time (t), and the rate constant for microbial inactivation (k):

$$\frac{dN}{dt} = -k\,N \tag{7.1}$$

Solving Eq. (7.1) using the initial condition that there are a known number of organisms at the beginning of the process ($N = N_0$ at $t = 0$), yields

$$\ln\left(\frac{N}{N_0}\right) = -k\,t \qquad (7.2)$$

Following the convention of the food industry, Eq. (7.2) is expressed in terms of common (base 10) logarithms instead of natural (base e) logarithms:

$$\log\left(\frac{N}{N_0}\right) = -\frac{k\,t}{2.303} \qquad (7.3)$$

After introducing the D value (the decimal reduction time, minutes), this becomes

$$\log\left(\frac{N}{N_0}\right) = -\frac{t}{D} \qquad (7.4)$$

or

$$\frac{N}{N_0} = 10^{-\frac{t}{D}} \qquad (7.5)$$

where $D = 2.303 / k$.

The z value is used to describe the relationship between two D values as a function of temperature:

$$\frac{D_{T_1}}{D_{T_2}} = 10^{\frac{T_2 - T_1}{z}} \qquad (7.6)$$

where z is a constant equal to the temperature difference required to achieve a log cycle change in D values. D_{T_1} and D_{T_2} are the decimal reduction times (in minutes) at T_1 and T_2, respectively. The D values decrease with increasing temperatures because microorganisms die faster at higher temperatures. Small z values mean the microbial death is very sensitive to temperature. Note, when the subscript *o* is used, it refers to a reference temperature of 250°F; hence, $D_o = D_{250°F}$.

An alternative approach for evaluating death kinetics of microorganisms is to use the Arrhenius equation to describe the influence of temperature on *k*. The "*z*-value" approach represents the common practice of the food industry. From a practical standpoint, both methods generally produce similar results when processing does not cover a wide range of temperatures. Typical *z* and *D* values for some organisms important in thermal processing are

- *Clostridium botulinum*: $D_{250°F} = 0.21$ min, $z =$ 18°F
- *Coxiellia burnettii*: $D_{152°F} = 0.32$ min, $z =$ 10.8°F
- P.A. 3679: $D_{250°F} = 0.80$ min, $z =$ 19.1°F
- *Salmonella*: $D_{140°F} = 0.30$ min, $z =$ 10°F

Many factors influence *D* and *z* values, so appropriate numbers must be identified for any particular product and target organism under consideration.

The sterilizing value (SV) of a process is defined in terms of the number of decimal reductions in the microbial population:

$$\text{SV} = \text{number of decimal reductions} = \log\left(\frac{N_0}{N}\right) \tag{7.7}$$

The SV achieved at a particular reference temperature (T_{ref}) can be calculated from the time of the process at T_{ref} divided by the appropriate *D* value:

$$\log\left(\frac{N_0}{N}\right) = \frac{t_{\text{at } T_{ref}}}{D^z_{T_{ref}}} \tag{7.8}$$

To ensure public health, thermal processes for low acid (pH greater than 4.6) foods are designed to achieve a minimum 12D reduction (SV=12) in *Clostridium botulinum* spores.

The target organism for milk pasteurization is *Coxiellia burnettii*. Assuming z = 10.8°F (6.0°C) and $D_{T=152.6^\circ F}^{z=10.8^\circ F} = 0.30\,\text{min}$, then the D value for this microorganism, at the pasteurization temperature (161°F = 71.7°C), is

$$D_{T=161^\circ F}^{z=10.8^\circ F} = (0.30)10^{\frac{152.6-161}{10.8}} = 0.005\text{ min} = 3.0\text{ s} \tag{7.9}$$

The Grade A Pasteurized Milk Ordinance (U.S. Food and Drug Administration) specifies a minimum process of 15 s at 161°F for milk; hence, given the above, the sterilizing value of the required process is

$$\text{SV} = \log\left(\frac{N_0}{N}\right) = \frac{t_{\text{at } T_{ref}}}{D_{T_{ref}}^{z}} = \frac{15}{3} = 5 \tag{7.10}$$

Rearranging Eq.(7.8), one can determine the time required to achieve a particular sterilizing value at different reference temperatures:

$$t_{\text{at } T_{ref}} = D_{T_{ref}}^{z} \log\left(\frac{N_0}{N}\right) \tag{7.11}$$

Using the above first order parameters for *Coxiellia burnettii*, and assuming a 5 log reduction as the target SV, yields an equation giving the equivalent process time (in seconds) at a particular temperature:

$$t_{\text{at } T_{ref}} = (0.30)(60)\left(10^{\frac{152.6-T}{10.8}}\right)(5) \tag{7.12}$$

Results are summarized in Table 7.1.

Table 7.1. Time to achieve a 5 log reduction in *Coxiellia burnettii* at different temperatures assuming $D_{T=152.6°F}^{z=10.8°F} = 0.32$ min = 18 s.

t, s	T, °F
157	150
53.9	155
18.6	160
6.4	165
2.2	170

Typical pasteurization requirements for various biological fluids are given in Table 7.2. A 5 log reduction in the microbial population can be expected in a typical pasteurization process such as that used for dairy products or apple cider. Minimum pasteurization requirements for some products, such as orange juice, are based on enzyme inactivation. Federal regulations for inactivation of *Salmonella* (Federal Register. February 27, 2001: 12589-12636) require a 6.5 log reduction in ground beef products and a 7.0 log reduction in poultry products.

7.2 The General Method

The general method is a numerical technique that accounts for the influence of different time-temperature combinations on the death of microorganisms. The thermal death time of a process (the F value) is, like D values, a function of z:

$$\frac{F_{T_1}}{F_{T_2}} = 10^{\frac{T_2 - T_1}{z}} \tag{7.13}$$

where F_{T_1} is thermal death time at T_1, and F_{T_2} is thermal death time at T_2. F values at different temperatures cause equivalent destruction of the target microorganisms.

Table 7.2. Typical pasteurization requirements for biological fluids.

Product	T (°F / °C)	Time
Apple Cider	154.6 / 68.1	14 s
Apple Cider	160 / 71.1	6 s
Milk	145 / 62.8	30 min
Milk	161 / 71.7	15 s
Milk	191 / 88.3	1.0 s
Cream, and milk chocolate	150 / 65.6	30 min
Cream	166 / 74.4	25 s
Ice Cream Mix	155 / 68.3	30 min
Ice Cream Mix	175 / 79.4	25 s
Ice Cream Mix	180 / 82.2	15 s
Orange Juice	194 / 90	60 s
Orange Juice	197.6 / 92	30 s
Orange Juice	208.4 / 98	6 s
Whole egg	140 / 60	3.5 min
Ultra-pasteurized (all dairy products)	280 / 137.8	2 s

Low Acid Fluids. Aseptically processed, low acid (pH greater than 4.6) foods are transported and thermally processed in fluid handling equipment. The common practice for characterizing the thermal treatment of these materials is to calculate the thermal death time at a reference temperature of 250°F (121.1°C) for an organism (*Clostridium botulinum*) with a z value of 18°F (10°C). The F-value at this temperature is defined, by convention, as $F_o = F_{250°F}$. Using the above equation, a relative kill time (RKT) can be calculated [Pflug, I.J. 1997. *J. Food Protection* 60(10): 1215-1223]:

$$RKT = \frac{F_T}{F_{250°F}} = 10^{\frac{250-T}{18}} = \frac{\text{min } at\ T}{\text{min } at\ 250°F} \tag{7.14}$$

Eq. (7.14) establishes (for low acid fluid foods) one minute at 250°F as a unit of lethality. If, for example, the product temperature is 230°F, then the RKT is

$$RKT = \frac{F_{230°F}}{F_{250°F}} = 10^{\frac{250-230}{18}} = 12.9\ \frac{\text{min } at\ 230°F}{\text{min } at\ 250°F} \tag{7.15}$$

meaning that holding a product at 230°F for 12.9 minutes has the equivalent lethal effect as holding the product at 250°F for 1 minute. It is convenient to introduce the lethal rate (LR) defined as the reciprocal of the relative kill time:

$$LR = \frac{1}{RKT} = 10^{\frac{T-250}{18}} = \frac{\text{min } at\ 250°F}{\text{min } at\ T} \tag{7.16}$$

In the above example,

$$LR = \frac{1}{RKT} = \frac{1}{12.9} = 0.077\ \frac{\text{min } at\ 250°F}{\text{min } at\ 230°F} \tag{7.17}$$

meaning that heating for 1 min at 230°F is equivalent to heating for 0.077 min at 250°F. In other words, heating for one minute at 230°F would add 0.077 min of lethality to a process. Using the lethal rate concept, the total lethality of a process can be calculated by adding up the cumulative lethality produced as product temperatures change during processing. Lethal rates, based on z = 18°F and a reference temperature of 250°F, are given in Table 7.3.

Table 7.3. Lethal rates at 250°F with z = 18°F.

T, °F	min at 250°F per min at T
210	0.006
215	0.011
220	0.022
225	0.041
230	0.077
235	0.147
240	0.278
245	0.527
250	1.000
255	1.896
260	3.594
265	6.813

From a public health standpoint, a thermal process for low acid foods must have a sterilizing value of 12 (i.e., a 12D process meaning SV = $\log(N_0/N) = 12$) with *Clostridium botulinum* ($D_{250°F}$ = 0.21 min and z = 18°F) as the target organism. Knowing this, the thermal death time at a reference temperature of 250°F may be calculated: $F_o = 12D = 12(0.21)$ = 2.52 min. Any process with F_o greater than 2.52 min is considered safe, and sometimes referred to as a "Bot Cook." F_o values of 5 min to

12 min are typically used because many spoilage organisms are more resistant to heat destruction than *Clostridium botulinum.*

Pasteurization of Dairy Products. A reference temperature of 250°F (121.1°C) is too high when using the general method to investigate fluid pasteurization. Considering the Pasteurized Milk Ordinance, reference temperatures of 161°F (71.7°C) for milk and 175°F (79.4°C) for ice cream mix, are more practical. In this work, the following lethality units will be established for pasteurization: 1 s at 161°F for milk, and 1 s at 175°F for ice cream mix. This yields the following relative kill times for milk and mix:

$$RKT_{milk} = \frac{F_T}{F_{161°F}} = 10^{\frac{161-T}{z}} = \frac{\text{sec } at\ T}{\text{sec } at\ 161°F} \quad (7.18)$$

$$RKT_{mix} = \frac{F_T}{F_{175°F}} = 10^{\frac{175-T}{z}} = \frac{\text{sec } at\ T}{\text{sec } at\ 175°F} \quad (7.19)$$

The value of z depends on the target organism for the process. For the pasteurization of dairy products where the target organism is *Coxiellia burnettii*, the organism responsible for the disease known as Q-fever, assuming z = 10.8°F (6.0°C) is reasonable. With this z value, the lethal rate equations for milk and ice cream mix can be established:

$$LR_{milk} = \frac{1}{RKT_{milk}} = 10^{\frac{T-161}{10.8}} = \frac{\text{sec } at\ 161°F}{\text{sec } at\ T} \quad (7.20)$$

$$LR_{mix} = \frac{1}{RKT_{mix}} = 10^{\frac{T-175}{10.8}} = \frac{\text{sec } at\ 175°F}{\text{sec } at\ T} \quad (7.21)$$

Lethal rates at 161°F and 175°F are summarized in Table 7.4. Alternative tables can be made, at various reference temperatures, using different z values. Lethal rates are less accurate as temperatures deviate from the reference temperature (T_{ref}).

Table 7.4. Lethal rates at 161°F and 175°F with z = 10.8°F.

Process Temperature, T (°F)	s at 161°F / s at T	s at 175°F / s at T
144	0.027	----
146	0.041	----
148	0.063	----
150	0.096	----
152	0.147	----
154	0.225	----
156	0.344	----
158	0.527	0.027
160	0.808	0.041
162	1.24	0.063
164	1.90	0.096
166	2.90	0.147
168	4.45	0.225
170	6.81	0.344
172	10.4	0.527
174	16.0	0.808
176	24.5	1.24
178	----	1.90
180	----	2.90
182	----	4.45
184	----	6.81
186	----	10.4
188	----	16.0
190	----	24.5

Process Lethality. The F-value of a process can be evaluated by combining the lethal rate (LR) and the time-temperature history:

$$F_{T_{ref}}^{z} = \int_{t_{start}}^{t_{end}} (LR)\,dt = \int_{t_{start}}^{t_{end}} \left(10^{\frac{T - T_{ref}}{z}} \right) dt \qquad (7.22)$$

In pasteurizing milk the F value (in this case, seconds at 161°F) of the process is calculated as

$$F_{T_{ref}}^{z} = F_{161^{\circ}F}^{10.8^{\circ}F} = \int_{t_{start}}^{t_{end}} (LR)\,dt = \int_{t_{start}}^{t_{end}} \left(10^{\frac{T - 161}{10.8}} \right) dt \qquad (7.23)$$

where t_{start} represents the time when the temperature is high enough to have a significant lethal effect. The superscript on F is used to clearly indicate the z value used in the analysis, and T_{ref} refers to the reference temperature for the process. If the lethal rate is an easily defined function of time (t), Eq (7.23) can be integrated analytically. Otherwise, the equation can be solved numerically by breaking the process into discrete units of time (Δt) and adding the thermal death occurring in each part of the process:

$$F_{T_{ref}}^{z} = \sum_{t_{start}}^{t_{end}} (LR)(\Delta t) \qquad (7.24)$$

Calculating the F-value of a process using the above equation is known as the "general method." The lethality of a process is the ratio of the F value of the process divided by the F value required for commercial sterility:

$$\text{Process Lethality} = \frac{\left(F_{T_{ref}}^{z} \right)_{process}}{\left(F_{T_{ref}}^{z} \right)_{required}} \qquad (7.25)$$

Process lethality must be greater than one to have an acceptable process. This concept is applied to milk pasteurization in Example Problem 8.11.

8 Example Problems

8.1 Comparison of Newtonian and Shear-Thinning Fluids

Honey is a Newtonian fluid and the temperature dependence of viscosity may be described with the Arrhenius equation (Eq. 1.16) using the appropriate constants:

$$\mu = A\exp\left(\frac{E_a}{R'T}\right) = 5.58\left(10^{-16}\right)\exp\left(\frac{10{,}972}{T}\right) \tag{8.1}$$

Blueberry pie filling at 20°C is a power law fluid with the following properties: $K = 6.0$ Pa s^n, $n = 0.43$. Examine the flow behavior of honey at 24°C and blueberry pie filling at 20°C, by comparing plots of shear stress versus shear rate, and apparent viscosity versus shear rate. Plot results over a shear rate range appropriate for pumping these fluids at 60 gpm in 3 inch (nominal diameter) tubing.

Solution. The internal diameter of 3 inch tubing is 2.834 inches = 0.07198 m (Appendix 9.3), and 60 gpm = 0.003785 m^{3}s^{-1}. The maximum shear rate (Eq. 1.10) for blueberry pie filling is calculated as

$$\dot{\gamma}_{max} = \left(\frac{3n+1}{4n}\right)\left(\frac{4Q}{\pi R^3}\right) = \left(\frac{3(.43)+1}{4(.43)}\right)\left(\frac{4(.003785)}{\pi(.07198/2)^3}\right) = 137.6 \text{ s}^{-1} \tag{8.2}$$

Assuming the minimum shear rate range is approximately 10% of the maximum; conducting the analysis over a range of approximately 10 to 140 s^{-1} is reasonable. Since $n = 1$ for honey, the correction factor is equal to 1.0 and maximum shear rate is 103.4 s^{-1}.

The viscosity of honey at 24°C is calculated from Eq. (8.1):

$$\mu = 5.58\left(10^{-16}\right)\exp\left(\frac{10,972}{24+273.15}\right) = 6.06 \text{ Pa s} \tag{8.3}$$

and, since honey is Newtonian,

$$\sigma = 6.06\,\dot{\gamma} \tag{8.4}$$

and the viscosity and the apparent viscosity are equal:

$$\mu = \eta = 6.06 \text{ Pa s} \tag{8.5}$$

The comparable equations for blueberry pie filling are

$$\sigma = K\left(\dot{\gamma}\right)^{n} = 6.0\left(\dot{\gamma}\right)^{0.43} \tag{8.6}$$

and

$$\eta = K\left(\dot{\gamma}\right)^{n-1} = 6.0\left(\dot{\gamma}\right)^{-0.57} \tag{8.7}$$

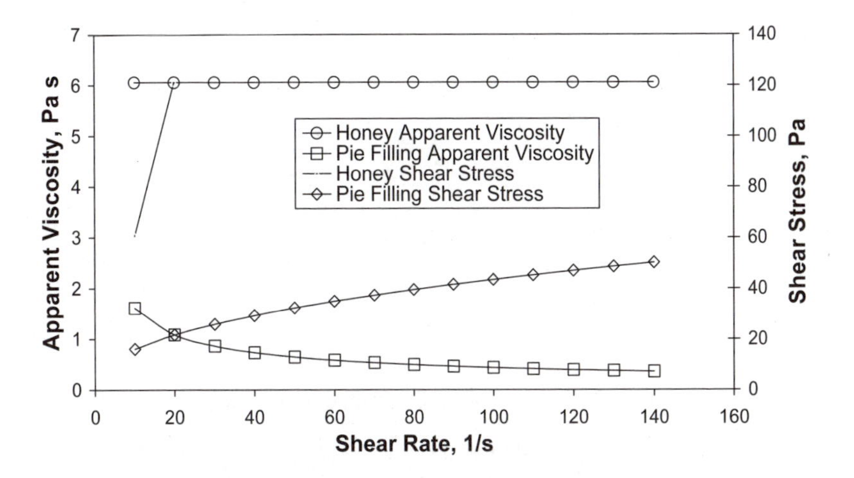

Figure 8.1. Apparent viscosity and shear stress of blueberry pie filling at 20°C (K = 6.0 Pa s^n, n = 0.43) and honey at 24°C (μ = 6.06 Pa s) over a shear rate range of 10 to 140 s^{-1}.

Plotting results (Fig. 8.1) illustrates the characteristic behavior of Newtonian and shear-thinning fluids. The shear stress versus shear rate curve for honey is a straight line (on the left of the figure) that increases rapidly from the origin. Due to the rapid increase in shear stress with shear rate, only two data points are shown; and since the shear stress becomes large so quickly, it is only plotted to a value of 20 s^{-1}. Since honey is Newtonian, the apparent viscosity is constant and appears as a horizontal line at $\eta = 6.06 \text{ Pa s}$. Blueberry pie filling is a typical shear-thinning fluid with a gradually rising shear stress curve, and a falling apparent viscosity curve.

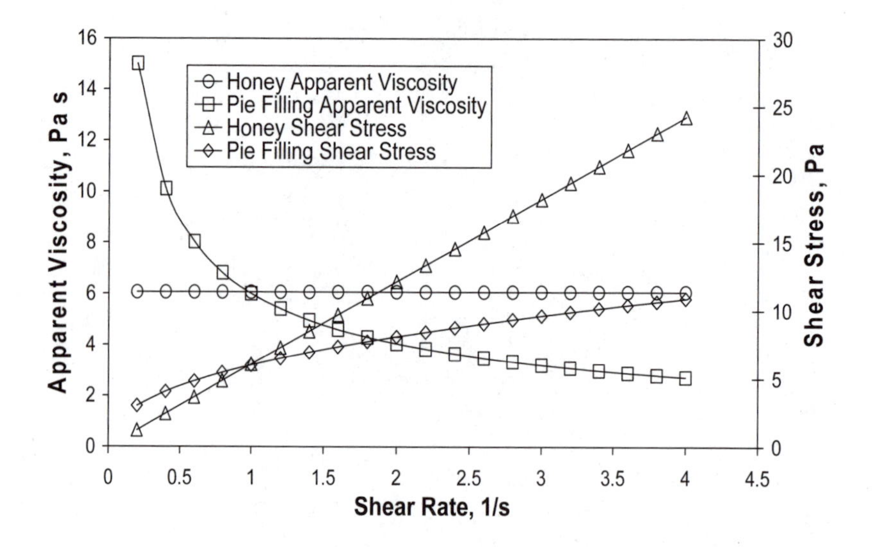

Figure 8.2. Apparent viscosity and shear stress of blueberry pie filling at 20°C ($K = 6.0$ Pa s^n, $n = 0.43$) and honey at 24°C ($\mu = 6.06$ Pa s) over a shear rate range of 0.2 to 4.0 s^{-1}.

It is also instructive to look at the flow behavior of these materials at low shear rates (Fig. 8.2). If extended to zero shear rate, both shear stress curves would go through the origin. Also, the shear stress curves intersect at 0.98 s^{-1} which make the apparent viscosities equal at that shear rate. It is interesting to note that one could mistakenly assume that blueberry pie filling was Newtonian and, by making a single point measurement at 1 s^{-1}, conclude that the rheological properties of honey and blueberry pie filling are equivalent. Mistakes like this can lead to large errors in the analysis of bioprocessing pipelines.

8.2 Herschel Bulkley and Casson Equations to Power Law

Rheological data for peach baby food were collected at 25°C over a shear rate range of approximately 10 to 300 s^{-1}. Data were summarized in terms of the Herschel-Bulkley equation:

$$\sigma = \sigma_o + K\dot{\gamma}^n \tag{8.8}$$

where σ_o = 13 Pa; K = 1.4 Pa s^n; n = 0.6. This equation is basically a power law expression with the yield stress (σ_o) representing the intercept found when the shear rate is equal to zero. Theoretically, the yield stress represents the minimum shear stress required to get the fluid moving. Due to the presence of the yield stress term in the Herschel-Bulkley equation, K given in that equation is not the same as K found in the standard power law equation: $\sigma = K\dot{\gamma}^n$.

The same peach baby food data were also evaluated in terms of the Casson model:

$$(\sigma)^{0.5} = K_1 + K_2(\dot{\gamma})^{0.5} \tag{8.9}$$

where K_1 = 3.7 Pa $^{0.5}$and K_2 = 0.22 $Pa^{0.5}$ $s^{0.5}$. This equation is commonly used to model the flow behavior of fluid chocolate. The Casson yield stress is the square of the intercept: $(K_1)^2$. In this particular case, the Casson yield stress equals 13.7 Pa.

To obtain pipeline design parameters (K and n), convert the above equations to the standard power law model, and compare results by plotting apparent viscosity (for both the original and converted equations) versus shear rate.

Solution. To solve this problem, the Herschel-Bulkley and Casson equations given above are used to generate "new" shear stress versus shear rate data sets (Table 8.1). After completing a logarithmic transformation of the data, power law parameters were determined by conducting a linear regression analysis of these data. Following this procedure, the Herschel-Bulkley curve converts to the power law curve as

$$\sigma = 7.36\left(\dot{\gamma}\right)^{0.35} \tag{8.10}$$

and the Casson curve converts as

$$\sigma = 7.59\left(\dot{\gamma}\right)^{0.34} \tag{8.11}$$

The corresponding apparent viscosity equations are, respectively,

$$\eta = 7.36\left(\dot{\gamma}\right)^{-0.65} \tag{8.12}$$

and

$$\eta = 7.59\left(\dot{\gamma}\right)^{-0.66} \tag{8.13}$$

Comparison of the apparent viscosity curves for the Casson case show very good agreement between the original and the converted equation (Fig. 8.3). The Herschel-Bulkley case produces similar results. From a practical standpoint, Eq. (8.10) and Eq. (8.11) are equivalent. Therefore, pipeline design problems could be solved using $n = 0.34$ and the average value of consistency coefficient: $K = 7.48$ Pa s^n.

Table 8.1. Shear stress-shear rate data set generated from the Herschel-Bulkley and Casson equations over the shear rate range of 10 to 300 s^{-1}.

$\dot{\gamma}$ (1/s)	$\sigma = 13 + 1.4(\dot{\gamma})^{0.6}$ (Pa)	$\sigma = \left(3.7 + 0.22(\dot{\gamma})^{0.5}\right)^2$ (Pa)
10	18.6	19.3
20	21.4	21.9
30	23.8	24.1
40	25.8	25.9
50	27.6	27.6
60	29.3	29.2
70	30.9	30.7
80	32.4	32.1
90	33.8	33.5
100	35.2	34.8
110	36.5	36.1
120	37.8	37.3
130	39.0	38.5
140	40.2	39.7
150	41.3	40.9
160	42.4	42.0
170	43.5	43.1
180	44.6	44.2
190	45.6	45.3
200	46.6	46.4
210	47.6	47.4
220	48.6	48.5
230	49.6	49.5
240	50.5	50.5
250	51.4	51.5
260	52.4	52.5
270	53.3	53.5
280	54.2	54.5
290	55.0	55.4
300	55.9	56.4

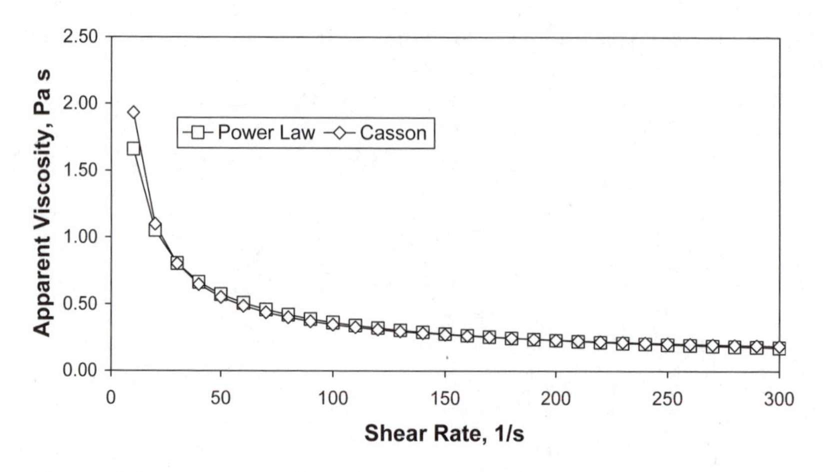

Figure 8.3. Apparent viscosity calculated from the original Casson equation and the power law equation.

8.3 Concentric Cylinder Data for Ice Cream Mix

Data for ice cream mix at two temperatures, representing pasteurization and aging conditions, were collected using a concentric cylinder viscometer (pictured in Fig. 2.2) with the following dimensions: R_c = 21.00 mm; R_b = 20.04 mm; h = 60.00 mm; h_o = 0.32 mm. Determine the rheological properties of this material at both temperatures. Compare rheograms, constructed using the equations for shear stress and the shear rate at the bob (Eq. 2.1 and Eq. 2.2), with those generated from average values (Eq. 2.3 and Eq. 2.4).

Table 8.2. Raw data for ice cream mix at 2°C and 83°C collected using a concentric cylinder viscometer: $R_c = 21.00$ mm; $R_b = 20.04$ mm; $h =$ 60.00 mm; $h_o = 0.32$ mm.

$T = 2$°C		$T = 83$°C	
Ω (rad/s)	M (N m)	Ω (rad/s)	M (N m)
2.08	0.01519	2.09	0.0019
2.61	0.01606	2.61	0.0021
3.13	0.01697	3.13	0.0022
3.64	0.01771	3.65	0.0024
4.15	0.01840	4.17	0.0025
4.70	0.01923	4.70	0.0026
5.23	0.02028	5.23	0.0026
6.26	0.02155	6.27	0.0030
7.30	0.02297	7.32	0.0031
8.36	0.02443	8.36	0.0033
9.40	0.02561	9.41	0.0036
10.44	0.02670	10.49	0.0038
11.49	0.02780	11.53	0.0040
12.54	0.02889	12.58	0.0042
13.62	0.03007	13.64	0.0045
14.68	0.03096	14.68	0.0047
15.71	0.03177	15.73	0.0048
16.76	0.03262	16.78	0.0050
17.81	0.03345	17.82	0.0053
18.86	0.03433	18.87	0.0055
19.91	0.03507	19.92	0.0055
20.97	0.03591	20.95	0.0057
22.01	0.03651	22.01	0.0059

Solution. Consider the data point at 83°C for an angular velocity of 10.49 rad s^{-1} in Table 8.2. Example calculations for the shear stress (Eq. 2.1) and shear rate at the bob (Eq. 2.2) are

$$\sigma_b = M\left[\frac{1}{2\pi R_b^2\left(h+h_o\right)}\right] = 0.0038\left[\frac{1}{2\pi(0.02004)^2(0.06032)}\right] = 25 \text{ Pa} \tag{8.14}$$

and, with $\alpha = R_c / R_b = 21.00/20.02 = 1.048$,

$$\dot{\gamma}_b = \Omega\left[\frac{2\alpha^2}{\alpha^2-1}\right] = 10.49\left[\frac{2(1.048)^2}{(1.048)^2-1}\right] = 234.8 \text{ s}^{-1} \tag{8.15}$$

The average values of shear stress (Eq 2.3) and shear rate (Eq. 2.4) are

$$\sigma_a = M\left[\frac{1+\alpha^2}{4\pi R_c^2\left(h+h_o\right)}\right] = 0.0038\left[\frac{1+(1.048)^2}{4\pi(0.021)^2(0.06032)}\right] = 23.9 \text{ Pa} \tag{8.16}$$

and

$$\dot{\gamma}_a = \Omega\left[\frac{\alpha^2+1}{\alpha^2-1}\right] = 10.49\left[\frac{(1.048)^2+1}{(1.048)^2-1}\right] = 224.3 \text{ s}^{-1} \tag{8.17}$$

These, along with values for the other data points, are found in Table 8.3 and plotted in Fig. 8.4. Regression analyses of the resulting curves yields the following rheological properties:

- bob equations at $T = 2$°C, $K = 21.10$ Pa s^n and $n = 0.39$;
- average equations at $T = 2$°C, $K = 20.52$ Pa s^n and $n = 0.39$;
- bob and average equations (curves were almost identical) at $T = 83$°C, $K = 1.76$ Pa s^n and $n = 0.49$.

Overall differences in calculations based on bob and average values are negligible.

Table 8.3. Values of shear stress and shear rate for ice cream mix at 83°C using equations reflecting behavior at the bob surface ($\dot{\gamma}_b$, σ_b) and average values between the bob and the cup ($\dot{\gamma}_a$, σ_a).

Ω (rad/s)	M (N m)	$\dot{\gamma}_b$ (1/s)	σ_b (Pa)	$\dot{\gamma}_a$ (1/s)	σ_a (Pa)
2.09	0.0019	46.8	12.5	44.7	11.9
2.61	0.0021	58.4	13.8	55.8	13.2
3.13	0.0022	70.1	14.5	66.9	13.8
3.65	0.0024	81.7	15.8	78.1	15.1
4.17	0.0025	93.4	16.4	89.2	15.7
4.70	0.0026	105.2	17.1	100.5	16.3
5.23	0.0026	117.1	17.1	111.9	16.3
6.27	0.0030	140.4	19.7	134.1	18.8
7.32	0.0031	163.9	20.4	156.6	19.5
8.36	0.0033	187.2	21.7	178.8	20.7
9.41	0.0036	210.7	23.7	201.2	22.6
10.49	0.0038	234.8	25.0	224.3	23.9
11.53	0.0040	258.1	26.3	246.6	25.1
12.58	0.0042	281.6	27.6	269.0	26.4
13.64	0.0045	305.4	29.6	291.7	28.2
14.68	0.0047	328.6	30.9	314.0	29.5
15.73	0.0048	352.1	31.5	336.4	30.1
16.78	0.0050	375.6	32.8	358.9	31.4
17.82	0.0053	398.9	34.8	381.1	33.3
18.87	0.0055	422.4	36.1	403.6	34.5
19.92	0.0055	445.9	36.1	426.0	34.5
20.95	0.0057	469.0	37.4	448.1	35.8
22.01	0.0059	492.7	38.8	470.7	37.0

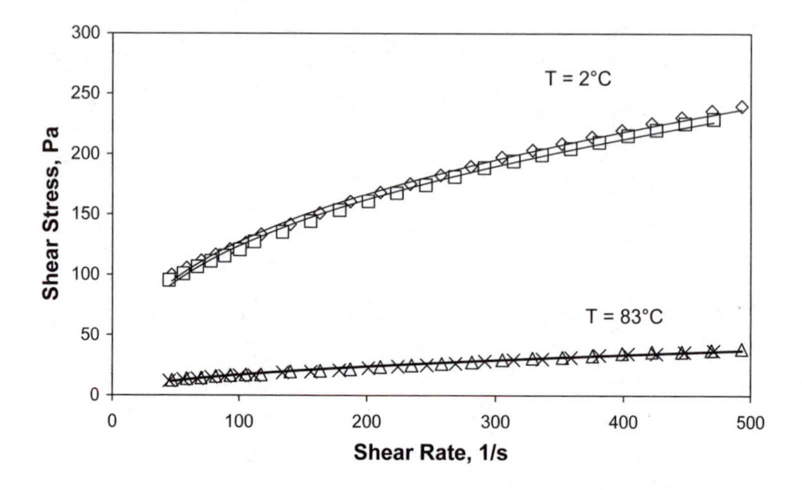

Figure 8.4. Rheogram for ice cream mix at 2°C and 83°C determined from values calculated at the surface of the bob (◊), and average or representative values (□).

8.4 Determination of the Mixer Coefficient

Data were collected for corn syrup at 23°C (μ = 2.3 Pa s) and honey at 23°C (μ = 7.5 Pa s) with a mixer viscometer using an interrupted helical screw impeller (shown in Fig. 2.4). Using raw data for corn syrup and honey (Table 8.4), determine the mixer coefficient (k'') for this system.

Solution. The product of viscosity and angular velocity ($\mu\ \Omega$) versus torque (M) is plotted in Fig. 8.5. The slope of the curve (k'') is determined by linear regression to be 3070.8 rad m^{-3}.

Table 8.4. Mixer viscometer data for two Newtonian fluids.

Corn Syrup at 23°C		Honey at 23°C	
M (μN m)	Ω (rad/s)	M (μN m)	Ω (rad/s)
722	1.04	1068	0.42
1454	2.07	3180	1.25
2930	4.14	4236	1.65
3660	5.17	6314	2.48
5146	7.24	7390	2.90
5860	8.26	9466	3.73
7310	10.4	10480	4.14
8030	11.4	12560	4.96
9510	13.4	13600	5.38
10260	14.5	15650	6.21
11760	16.6	16640	6.62
12510	17.6	18620	7.43
14060	19.7	19660	7.86
14840	20.7	21720	8.70
16390	22.8	22650	9.09
17220	23.8	24680	9.94
18870	25.9	25680	10.4
19660	26.9	27580	11.2
21250	29.0	28660	11.6

8.5 Mixer Viscometry Data for Pasta Sauce

Using the interrupted helical screw impeller illustrated in Fig. 2.4, data (Table 8.5) were collected for Prego Fresh Mushrooms Pasta Sauce (Campbell Soup Co., Camden, NJ). Given $k'' = 3047.4$ rad m^{-3} and $k' = 1.6$ rad^{-1} for this impeller, plot apparent viscosity versus shear rate, and determine the power law fluid properties (K and n) of the pasta sauce. This fluid could not be evaluated in a conventional concentric cylinder viscometer because of the large particles present in the material: tomato and mushroom pieces with a single dimension as large as 1.48 and 0.81 cm, respectively, were measured.

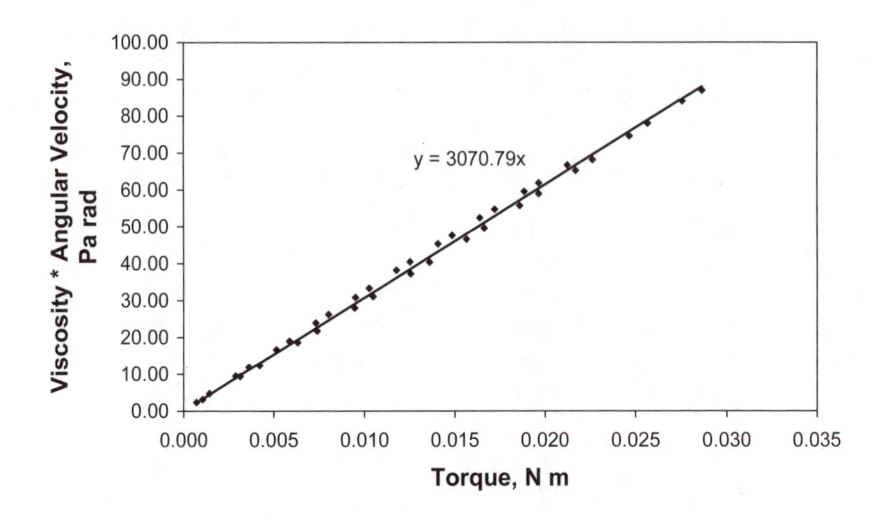

Figure 8.5. Plot of $\mu\Omega$ versus M for corn syrup and honey at 83°C to find k'' : the slope of the plot = $k'' = 3070.8$ rad m^{-3}.

Table 8.5. Mixer viscometer data, collected with the helical screw impeller, for Prego pasta sauce (with mushrooms) at 80°C.

Ω (rad/s)	M (μN m)	Ω (rad/s)	M (μN m)
2.99	5420	11.8	8080
4.17	5870	13.0	7900
4.75	6020	13.6	7890
5.93	6290	14.7	8030
6.51	6290	15.3	8160
7.70	6920	16.5	8290
8.26	7170	17.1	8350
9.44	7400	18.3	8440
10.0	7590	18.8	8580
11.2	7610	20.0	8940

Solution. To solve this problem, first consider the calculations involved for one data point. At an angular velocity of 10 rad s^{-1}, the apparent viscosity (Eq. 2.19) is calculated as

$$\eta = \frac{M\,k''}{\Omega} = \frac{7590\left(10^{-6}\right)3047.4}{10} = 2.31 \text{ Pa s} \tag{8.18}$$

and the average shear rate (Eq. 2.12) is

$$\dot{\gamma}_a = k'\Omega = 1.6\left(10\right) = 16 \text{ s}^{-1} \tag{8.19}$$

Plotting these results (Fig. 8.6), along with the other values determined from the raw data given in Table 8.4, shows the non-linear relationship between the variables. The shear-thinning properties are found by fitting a trend line to the data: $K = 17.5$ Pa s^n, and $n = 0.26$.

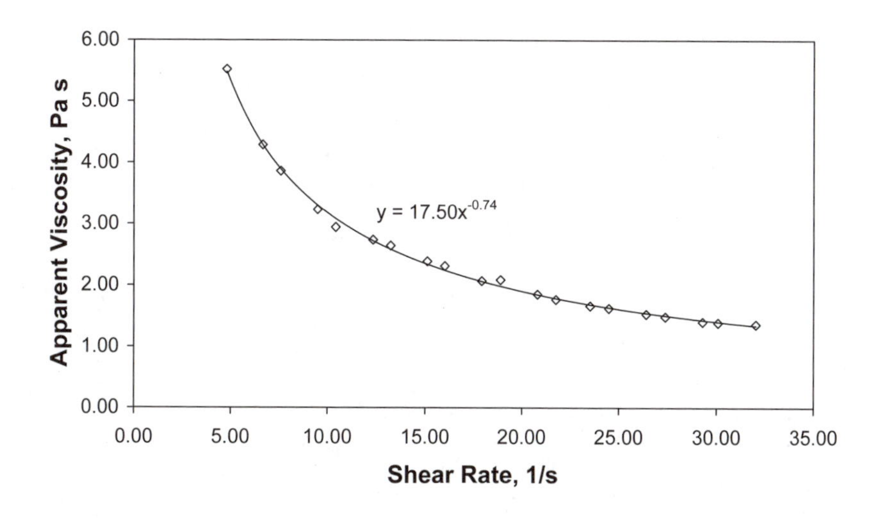

Figure 8.6. Apparent viscosity versus shear rate for pasta sauce at 80°C with regression analysis showing $K = 17.50$ Pa s^n, and $n - 1 = -0.74$

8.6 Calculating Pressure Drop with Effective Viscosity

Consider pumping a thick, shear-thinning tomato sauce product ($K = 30$ Pa s^n, $n = 0.2$, $\rho = 1010$ kg m^{-3}) flowing at 40 gpm over a 50 m length of 3 inch nominal diameter (internal diameter = 0.072 m) pipe. Determine the ratio of the actual pressure drop over the pipe calculated using standard fluid mechanics to the pressure drop predicted using the effective viscosity technique. How much error is introduced using effective viscosity?

Solution. The volumetric flow rate and mean velocity (Eq. 4.2) are calculated from the given parameters:

$$Q = 40 \text{ gpm} = \frac{40}{60(264.17)} = 0.00252 \text{ m}^3\text{s}^{-1} \quad (8.20)$$

$$\bar{u} = \frac{Q}{A} = \frac{0.00252}{(\pi/4)(0.072)^2} = 0.619 \text{ m s}^{-1} \quad (8.21)$$

The power law Reynolds number (Eq. 4.9) and the actual Fanning friction factor (found using Eq. 4.13 for laminar flow) are found as

$$N_{\text{Re},PL} = \left(\frac{D^n (\bar{u})^{2-n} \rho}{8^{n-1} K}\right)\left(\frac{4n}{3n+1}\right)^n$$

$$= \left(\frac{(0.072)^{0.2} (0.619)^{2-0.2} 1010}{8^{0.2-1}(30)}\right)\left(\frac{4(0.2)}{3(0.2)+1}\right)^{0.2} \quad (8.22)$$

$$= 38.5$$

$$f = \frac{16}{N_{\text{Re},PL}} = \frac{16}{38.5} = 0.415 \quad (8.23)$$

There is no standard method to calculate effective viscosity: It is usually determined as the value of the apparent viscosity calculated at the maximum shear rate in the pipe by assuming the fluid is Newtonian (Eq. 1.10 with $n = 1$):

$$\dot{\gamma}_{\max} = \frac{4Q}{\pi R^3} = \frac{4(0.00252)}{\pi\left(\frac{0.072}{2}\right)^3} = 68.9 \text{ s}^{-1} \tag{8.24}$$

Using this shear rate, and the definition of apparent viscosity, the effective viscosity can be calculated:

$$\mu_{\text{effective}} = K(\dot{\gamma})^{n-1} = 30(68.9)^{0.2-1} = 1.02 \text{ Pa s} \tag{8.25}$$

Now, assuming Newtonian behavior, values of the Reynolds number and the effective Fanning friction factor can be found:

$$N_{\text{Re}} = \frac{\rho D \overline{u}}{\mu_{\text{effective}}} = \frac{1010(0.072)(0.619)}{1.02} = 44.1 \tag{8.26}$$

$$f_{\text{effective}} = \frac{16}{N_{\text{Re}}} = \frac{16}{44.1} = 0.362 \tag{8.27}$$

The pressure loss in the pipe is calculated by application of the mechanical energy balance equation (Eq. 3.2):

$$\Delta P = \frac{2\rho f \overline{u}^2 L}{D} \tag{8.28}$$

Substituting the appropriate numbers, actual pressure loss over the pipe is

$$\Delta P_{\text{actual}} = \frac{2(1010)(0.415)(0.619)(50)}{0.072} = 223 \text{ kPa} \tag{8.29}$$

The ratio of the predicted (using the effective viscosity idea) over the actual pressure drop is

$$\frac{\Delta P_{predicted}}{\Delta P} = \frac{f_{effective}}{f} = \frac{0.362}{0.415} = 0.87 \tag{8.30}$$

Eq. (8.30) shows the predicted value is 87% of the actual value, meaning, in this particular example, there is a 13% error using the effective viscosity method to predict pressure drop. Depending on the value of the flow behavior index, and the choice of shear rates, the pressure drop predicted using effective viscosity can be significantly different than the true pressure drop.

It is instructive to compare the effective viscosity and the apparent viscosity. Actual shear rates in the tube range from zero at the center of the tube to the maximum value at the wall that must be calculated (Eq. 1.10) with due consideration of the flow behavior index:

$$\dot{\gamma}_{max} = \left(\frac{3n+1}{4n}\right)\left(\frac{4Q}{\pi R^3}\right) = \left(\frac{3(0.2)+1}{4(0.2)}\right)\left(\frac{4(0.00252)}{\pi\left(\frac{0.072}{2}\right)^3}\right) = 137.8 \text{ s}^{-1} \tag{8.31}$$

At this shear rate, the apparent viscosity is at the minimum value because there is a viscosity distribution at a given flow rate:

$$\eta_{min} = K(\dot{\gamma})^{n-1} = 30(137.8)^{0.2-1} = 0.583 \text{ Pa s} \tag{8.32}$$

At a shear rate equal to 1.0 s^{-1}, the apparent viscosity is 30 Pa s. If the flow behavior index was equal to 1.0, this would be the Newtonian viscosity.

Since the apparent viscosity is a function of the shear rate, and the shear rate is a function of the radius, η can also be written in terms of the radius:

$$\eta = K\left[\left(\frac{3n+1}{4n}\right)\left(\frac{4Q}{\pi R^3}\right)\left(\frac{r}{R}\right)^{1/n}\right]^{n-1} \tag{8.33}$$

As the radius goes to zero at the center of the tube, the apparent viscosity approaches infinity. Hence, in this particular example problem, one can find apparent viscosities in the pipe ranging from infinity to 0.583 Pa s. This illustrates why arbitrarily matching an effective viscosity to an apparent viscosity can lead to significant errors in predicting pipeline pressure losses. It also highlights the point that one parameter is inadequate to describe the rheological behavior of a non-Newtonian fluid; at least two (in this case K and n) are needed.

8.7 Generating a System Curve for Pumping Cream

Assume a fluid unloading system (tanker truck to storage tank) has been designed with the following considerations:

1) The change in elevation when the truck is nearly empty and the storage tank (with pipe entrance in the bottom of tank) is nearly full (giving the maximum potential energy difference) is 3.5 m.
2) The fluid involved is cream at 5°C: $\rho = 985$ kg m^{-3} and $\mu = 45$ cP. Both tanks are vented to the atmosphere during unloading.
3) The processing system includes the following standard pipeline components in the suction line:
   - entrance to 3 inch tubing,
   - 2.5 m of 3 inch tubing (from Appendix 9.3, internal diameter = 2.83 inch = 0.072 m),
   - three 90° elbows (standard, welded) in the 3 inch line;

   and the following standard components in the discharge line:
   - 19.0 m of 2.5 inch tubing (from Appendix 9.3, internal diameter = 2.37 inch = 0.0602 m),
   - seven 90° elbows (standard, welded) in the 2.5 inch line,
   - 2.5 inch tee (standard, welded) used as elbow,
   - exit from 2.5 inch tubing.

4) The system will have one strainer, and two pneumatic valves, in the 2.5 inch discharge line. Experimental data for these components, collected for water at room temperature ($\rho = 998$ kg m^{-3}), are provided by the manufacturer as:

Q (gpm)	ΔP, Strainer (Pa)	ΔP, Valve (Pa)
30	500	483
40	1000	966
50	1500	1449
60	2000	1933
70	2500	2416

Provide the following:

1) The symbolic (theoretical) equations for the work input to the pump showing all elements of the system to be evaluated; and the calculation for total system head.

2) Detailed numerical calculations for a flow rate of 50 gpm. Include a calculation of the hydraulic power requirement.

3) The system curve required to select a centrifugal pump for this processing system. Assume the flow rate range under consideration is 6.8 to 15.9 m^3 hr^{-1} (30 – 70 gpm). Summarize results in a plot of total system head (m) versus flow rate (m^3/ hr).

Solution, Part 1. Since both tanks are vented to the atmosphere, the pressures are equal so $P_2 - P_1 = 0$; and the velocity at the liquid level in the tank is very small making the assumption that $\bar{u}_2 = \bar{u}_1 = 0$ acceptable. Taking these facts into consideration, the work input to the pump may be calculated from the mechanical energy balance equation (Eq. 3.2) as

$$W = g(z_2 - z_1) + \sum F \tag{8.34}$$

where:

$$\sum F = \sum \frac{k_f \bar{u}^2}{2} + \sum \frac{2f\bar{u}^2 L}{D} + \sum \left(\frac{\Delta P}{\rho}\right)_{\text{equipment}} \tag{8.35}$$

The summation of friction loss must be expanded to include all the specific elements of the system (using Eq. 5.4 and 5.5 to find k_f values):

$$\sum \frac{k_f \bar{u}^2}{2} = \left[\frac{\bar{u}^2}{2}\left(k_{f,\text{ entrance}} + 3k_{f,\text{ elbow}}\right)\right]_{3\text{ inch line}} + \left[\frac{\bar{u}^2}{2}\left(k_{f,\text{ tee}} + 7k_{f,\text{ elbow}} + k_{f,\text{ exit}}\right)\right]_{2.5\text{ inch line}} \tag{8.36}$$

$$k_{f,\text{ entrance}} = \frac{160}{N_{\text{Re}}} + 0.50 \tag{8.37}$$

$$k_{f,elbow} = \frac{800}{N_{\text{Re}}} + 0.25\left(1 + \frac{1}{D_{inch}}\right) \tag{8.38}$$

$$k_{f,tee} = \frac{800}{N_{\text{Re}}} + 0.80\left(1 + \frac{1}{D_{inch}}\right) \tag{8.39}$$

$$k_{f,\text{ exit}} = 1.0 \quad \text{(from Table 5.3)} \tag{8.40}$$

$$\sum \frac{2f\bar{u}^2 L}{D} = \left[\frac{2f\bar{u}^2 L}{D}\right]_{3\text{ inch line}} + \left[\frac{2f\bar{u}^2 L}{D}\right]_{2.5\text{ inch line}} \tag{8.41}$$

$$\sum\left(\frac{\Delta P}{\rho}\right)_{\text{equipment}}=\left(\frac{f}{f_{\text{water}}}\right)\left(\frac{\Delta P}{\rho}\right)_{\text{strainer, water}}+2\left(\frac{f}{f_{\text{water}}}\right)\left(\frac{\Delta P}{\rho}\right)_{\text{pneumatic valve, water}} \tag{8.42}$$

After the total work input (W) has been calculated, the total system head may be determined by considering the gravitational acceleration:

$$H_s=\frac{W}{g} \tag{8.43}$$

Solution, Part 2. Start by establishing some preliminary information: cream properties are $\mu = 45$ cP = 0.045 Pa s and $\rho = 985$ kg m^{-3}; $Q = 50$ gpm = 11.4 m^3 hr^{-1} = 0.00315 m^3 s^{-1}; $D = 0.072$ m, and $L = 2.5$ m; and

$$\bar{u}=\frac{4Q}{\pi D^2}=\frac{4(0.00315)}{\pi(0.072)^2}=0.774\,\text{m}\,\text{s}^{-1} \tag{8.44}$$

$$N_{\text{Re}}=\frac{D\bar{u}\rho}{\mu}=\frac{0.072(0.774)(985)}{0.045}=1220 \tag{8.45}$$

$$k_{f,\text{ entrance}}=\frac{160}{1220}+0.50=0.63 \tag{8.46}$$

$$k_{f,\text{ elbow}}=\frac{800}{1220}+0.25\left(1+\frac{1}{2.83}\right)=0.99 \tag{8.47}$$

$$f=\frac{16}{N_{\text{Re}}}=\frac{16}{1220}=0.0131 \tag{8.48}$$

In the 2.5 inch section of tubing, $D = 0.0602$ m, and $L = 19.0$ m; $k_{f,\text{ exit}}=1.0$; and

$$\bar{u} = \frac{4Q}{\pi D^2} = \frac{4(0.00315)}{\pi(0.0602)^2} = 1.107\,\mathrm{m\,s^{-1}} \tag{8.49}$$

$$N_{\mathrm{Re}} = \frac{D\bar{u}\rho}{\mu} = \frac{0.0602(1.107)(985)}{0.045} = 1459 \tag{8.50}$$

$$k_{f,\,\mathrm{tee}} = \frac{800}{1459} + 0.80\left(1 + \frac{1}{2.37}\right) = 1.69 \tag{8.51}$$

$$k_{f,\,\mathrm{elbow}} = \frac{800}{1459} + 0.25\left(1 + \frac{1}{2.37}\right) = 0.90 \tag{8.52}$$

$$f = \frac{16}{N_{\mathrm{Re}}} = \frac{16}{1459} = 0.0110 \tag{8.53}$$

The properties of water and steam (Appendix 9.5 and 9.7) at room temperature are needed to determine the Reynolds number at the given volumetric flow rate: $\mu = 0.001$ Pa s; $\rho = 998$ kg m^{-3}. These data are needed to find the friction factor for water required in the equipment friction loss correction. The Reynolds number for water in the 2.5 inch tube is

$$N_{\mathrm{Re}} = \frac{D\bar{u}\rho}{\mu} = \frac{(0.0602)(1.107)(998)}{0.001} = 66508 \tag{8.54}$$

Calculating the friction factor with the Blasius equation (Eq. 4.4) yields

$$f_{water} = \frac{0.0791}{N_{\mathrm{Re}}^{0.25}} = \frac{0.0791}{(66508)^{0.25}} = 0.00493 \tag{8.55}$$

Slightly different values of the friction factor are found using other equations: f = 0.00485 using the Haaland equation (Eq. 4.5), and f = 0.00487 using the Churchill equation (Eq. 4.6). These differences are negligible.

Now, the overall equations can be evaluated. The friction loss in the straight is

$$\sum \frac{2f\bar{u}^2L}{D} = \left[\frac{2f\bar{u}^2L}{D}\right]_{3\text{ inch line}} + \left[\frac{2f\bar{u}^2L}{D}\right]_{2.5\text{ inch line}}$$
$$= \left[\frac{2(0.0131)(0.774)^2(2.5)}{0.0720}\right] + \left[\frac{2(0.0110)(1.107)^2(19)}{0.0602}\right] = 9.03\text{ J kg}^{-1} \tag{8.56}$$

and the combined loss for the fittings and the exit is

$$\sum \frac{k_f\bar{u}^2}{2} = \left[\frac{\bar{u}^2}{2}\left(k_{f,\text{ entrance}} + 3k_{f,\text{ elbow}}\right)\right]_{3\text{ inch line}}$$
$$+ \left[\frac{\bar{u}^2}{2}\left(k_{f,\text{ tee}} + 7k_{f,\text{ elbow}} + k_{f,\text{ exit}}\right)\right]_{2.5\text{ inch line}}$$
$$= \left[\frac{(0.774)^2}{2}(0.63 + 3(0.99))\right]$$
$$+ \left[\frac{(1.107)^2}{2}(1.69 + 7(0.90) + 1.0)\right]$$
$$= 6.59\text{ J kg}^{-1} \tag{8.57}$$

and the loss in the strainer and two pneumatic valves is

$$\sum \left(\frac{\Delta P}{\rho}\right)_{\text{equipment}} = \left(\frac{0.0110}{0.00493}\right)\left(\frac{1500}{998}\right) + 2\left(\frac{0.0110}{0.00493}\right)\left(\frac{1449}{998}\right) = 9.83\text{ J kg}^{-1} \tag{8.58}$$

Total friction loss is the sum of the above component losses:

$$\sum F = 9.03 + 6.59 + 9.83 = 25.45\text{ J kg}^{-1} \tag{8.59}$$

The total work input required by the pump, Eq. (8.34), can be determined by including the potential energy difference:

$$W = 9.81(3.5) + 25.45 = 59.79\text{ J kg}^{-1} \tag{8.60}$$

Note that more than 40% of the energy input is used to overcome friction losses in the pipeline, and the rest is needed to lift the fluid. Using the results from Eq.(8.60), the system head at 50 gpm (11.4 $m^3\ hr^{-1}$) can be found using Eq. (8.43):

$$H_s = \frac{W}{g} = \frac{59.79}{9.81} = 6.09 \text{ m} \tag{8.61}$$

Also, the pressure increase over the pump at the 50 gpm flow rate can be calculated using the total work input:

$$\Delta P = W\rho = 59.79(985) = 58.9 \text{ kPa } (8.5 \text{ psi}) \tag{8.62}$$

The hydraulic power (Eq. 3.14) requirement at 50 gpm is

$$\Phi = W\dot{m} = WQ\rho = 59.79(0.00315)(985) = 186 \text{ W} \tag{8.63}$$

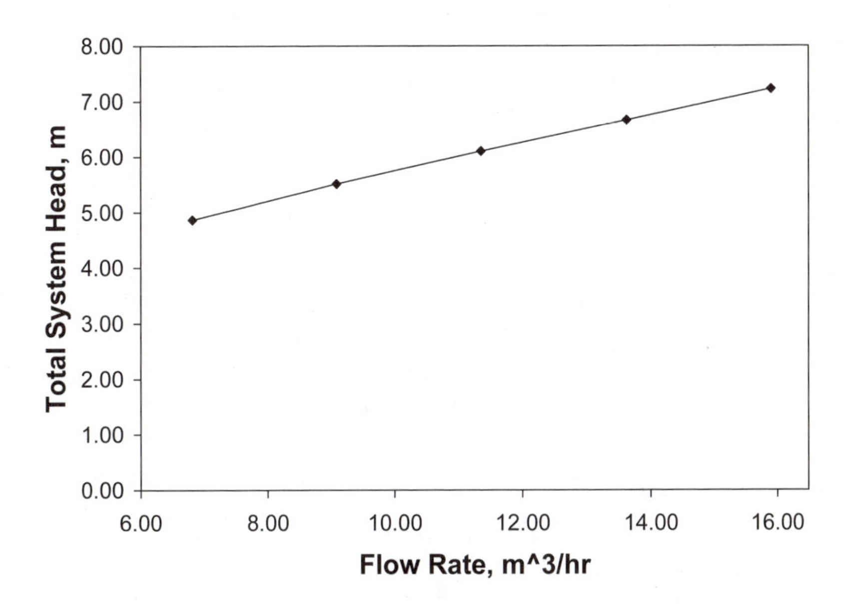

Figure 8.7. System curve for cream at 5°C.

Solution, Part 3. Recalculating the above calculations, Eq. (8.44) through Eq. (8.61), at different flow rates; and plotting the results yields the system curve for the process (Fig. 8.7). The intersection of the pump curve (for a centrifugal pump) and the system curve would establish the cream flow rate through the system.

8.8 Positive Displacement Pump for Pulpy Fruit Juice

Assume the same physical elements in the system as Example Problem 8.7 with the addition of a 3 inch to 2.5 inch contraction at the pump exit. The system was designed to transport 110 gpm pulpy fruit juice (a shear-thinning fluid with K = 0.43 Pa s^n, n = 0.6, ρ = 1030 kg m^{-3}) at 10°C using a rotary lobe type positive displacement pump. The following items are located before the pump (in the suction line):

- entrance to 3 inch tubing,
- 2.5 m of 3 inch tubing (from Appendix 9.3, internal diameter = 2.83 inch = 0.072 m),
- three 90° elbows (standard, welded) in the 3 inch line.

The remaining items (everything in the 2.5 inch line) are found after the pump (in the discharge line):

- 3 inch to 2.5 inch contraction (θ= 9.5°),
- 19.0 m of 2.5 inch tubing (from Appendix 9.3, internal diameter = 2.37 inch = 0.0602 m),
- seven 90° elbows (standard, welded) in the 2.5 inch line,
- 2.5 inch tee (standard, welded) used as elbow,
- exit from 2.5 inch tubing,
- one strainer, and two pneumatic valves.

Assume the elevation difference (3.5 m) remains unchanged. Also, literature from the manufacturer indicates that the pressure drop (for water at 110 gpm) over the strainer and each pneumatic valve can be expected to be 4000 Pa and 3030 Pa, respectively.

Determine the symbolic (theoretical) solutions for the following:

1) Work input for the pump with an explanation of what factors in the solution are different from those found in the solution for the Newtonian fluid examined in Example Problem 8.7.
2) Pressure at the pump inlet, and the $(NPSH)_A$.
3) Pressure at the pump outlet.

Using the above theoretical solutions, find the numerical values for:

4) The total energy input (J/kg), and the total system head (m) for this design.
5) The absolute pressure (kPa and psia) at the inlet of the pump (assume zero elevation difference, $z_1 = z_2$); and the $(NPSH)_A$.
6) The absolute pressure (kPa and psia) at the outlet of the pump assuming the maximum potential energy difference: outlet liquid level is 3.5 m above the pump.
7) The total pressure drop (in kPa and psi) across the pump.
8) The hydraulic power requirement for the system (Watts and hp).

Theoretical Solution:

1) Solution for W is the same as that found for cream in Example Problem 8.7 except that different numerical values for $\overline{u}$, and f are needed; $N_{\text{Re},PL}$ is used in place of N_{Re} in calculating the friction loss coefficients; and the friction loss for the 3 inch to 2.5 inch contraction must be added.

2) Solution for pump entrance pressure (P_2, absolute pressure) with point one located at the fluid level of the system entrance is found using the mechanical energy balance equation (Eq. 3.1) and recognizing that $W = \bar{u}_1 = 0$; and $z_1 = z_2$:

$$(P_2)_{\text{pump entrance}} = \rho\left(\frac{P_1}{\rho} - \frac{\bar{u}_2^2}{\alpha} - \sum F\right) \tag{8.64}$$

where $P_1 = 1$ atm, absolute pressure; and

$$\sum F = \sum \frac{k_f \bar{u}^2}{2} + \sum \frac{2 f \bar{u}^2 L}{D} \tag{8.65}$$

and

$$\sum \frac{k_f \bar{u}^2}{2} = \left[\frac{\bar{u}^2}{2}\left(k_{f,\text{ entrance}} + 3k_{f,\text{ elbow}}\right)\right]_{\text{3 inch line}} \tag{8.66}$$

$$\sum \frac{2 f \bar{u}^2 L}{D} = \left[\frac{2 f \bar{u}^2 L}{D}\right]_{\text{3 inch line}} \tag{8.67}$$

$$\alpha = \frac{2(2n+1)(5n+3)}{3(3n+1)^2} \tag{8.68}$$

Use the value of the pump entrance pressure given above, and the absolute vapor pressure of water at the processing temperature (P_v) to determine the net positive suction head available (Eq. 3.24):

$$(NPSH)_A = \frac{(P_2)_{\text{pump entrance}} - P_v}{g\rho} \tag{8.69}$$

3) The solution for pump exit pressure can be found with the known values of the pump entrance pressure and the work input provided by the pump:

$$P_{\text{pump exit}} = W\rho + P_{\text{pump entrance}} \tag{8.70}$$

Numerical Solution. Start by establishing some preliminary information: pulpy fruit juice properties at 10°C are $K = 0.43$ Pa s^n, $n = 0.6$, and $\rho = 1030$ kg/m^{-3}; $Q = 110$ gpm $= 0.00694$ m^3 s^{-1}; mass flow rate $= 7.15$ kg s^{-1}; $D = 0.072$ m, and $L = 2.5$ m; $\bar{u} = 1.70$ m s^{-1}; and the relevant equations for the inlet side of the pump are

$$N_{\text{Re},PL} = \left(\frac{(0.072)^{0.6}(1.70)^{2-0.6}1030}{8^{0.6-1}(0.43)}\right)\left(\frac{4(0.6)}{3(0.6)+1}\right)^{0.6} = 2183 \tag{8.71}$$

$$\left(N_{\text{Re},PL}\right)_{critical} = 2100 + 875(1-n) = 2100 + 875(1-0.6) = 2450 \tag{8.72}$$

$$k_{f,\text{ entrance}} = \frac{160}{2183} + 0.50 = 0.57 \tag{8.73}$$

$$k_{f,\text{ elbow}} = \frac{800}{2183} + 0.25\left(1 + \frac{1}{2.87}\right) = 0.70 \tag{8.74}$$

$$f = \frac{16}{2183} = 0.00733 \text{ (since flow is laminar)} \tag{8.75}$$

In the 2.5 inch section, $D = 0.0602$ m and $L = 19.0$ m; $\bar{u} = 2.44$ m s^{-1}; $N_{\text{Re},PL} = 3237$ (indicating transitional flow); $k_{f,\text{ exit}} = 1.0$; and

$$k_{f,\text{ tee}} = \frac{800}{3237} + 0.80\left(1 + \frac{1}{2.37}\right) = 1.38 \tag{8.76}$$

$$k_{f,\text{ elbow}} = \frac{800}{3237} + 0.25\left(1 + \frac{1}{2.37}\right) = 0.60 \tag{8.77}$$

$$f = 0.00732 \text{ (from Appendix 9.11)} \tag{8.78}$$

The friction loss coefficient for the contraction is found with the following equation (Table 5.7) for $0° < \theta < 45°$ and $N_{Re} \leq 2500$:

$$k_{f,\text{contraction}} = \left[1.2 + \frac{160}{2183}\right]\left[\left(\frac{72.0}{60.2}\right)^4 - 1\right]\left[1.6\sin\left(\frac{9.5}{2}\right)\right] = 0.18 \quad (8.79)$$

The properties of water (Appendix 9.5 and 9.7) at room temperature are needed to determine the Reynolds number at the given volumetric flow rate: $\mu = 0.001$ Pa s; $\rho = 998$ kg m^{-3}. These data are used to find the friction factor for water required in the equipment friction loss correction. The Reynolds number for water in the 2.5 inch tube is:

$$N_{Re} = \frac{D\bar{u}\rho}{\mu} = \frac{(0.0602)(2.44)(998.2)}{0.001} = 146{,}623 \quad (8.80)$$

Calculating the friction factor for water with the Churchill equation (Eq. 4.6), yields

$$f_{water} = 0.00413 \quad (8.81)$$

4) Finding the work input to the pump. Using the above information, the overall equations can be evaluated. The friction loss in the straight pipe is

$$\sum \frac{2f\bar{u}^2L}{D} = \left[\frac{2f\bar{u}^2L}{D}\right]_{3\text{ inch line}} + \left[\frac{2f\bar{u}^2L}{D}\right]_{2.5\text{ inch line}}$$
$$= \left[\frac{2(0.00733)(1.70)^2(2.5)}{0.0720}\right] + \left[\frac{2(0.00732)(2.44)^2(19)}{0.0602}\right] = 29.0 \text{ J kg}^{-1} \quad (8.82)$$

and the combined loss in the fittings and the exit is

$$\sum \frac{k_f \overline{u}^2}{2} = \left[\frac{\overline{u}^2}{2} \left(k_{f,entrance} + 3k_{f,elbow} \right) \right]_{3 \text{ inch line}}$$
$$+ \left[\frac{\overline{u}^2}{2} \left(k_{f,contraction} + k_{f,tee} + 7k_{f,elbow} + k_{f,exit} \right) \right]_{2.5 \text{ inch line}}$$
$$= \left[\frac{(1.70)^2}{2} \left(0.57 + 3(0.70) \right) \right]$$
$$+ \left[\frac{(2.44)^2}{2} \left(0.18 + 1.38 + 7(0.60) + 1.0 \right) \right]_{2.5 \text{ inch line}}$$
$$= 24.1 \text{ J kg}^{-1} \tag{8.83}$$

and the loss in the strainer, and two pneumatic valves, is

$$\sum \left(\frac{\Delta P}{\rho} \right)_{\text{equipment}} = \left(\frac{0.00732}{0.00413} \right) \left(\frac{4000}{998} \right) + 2 \left(\frac{0.00732}{0.00413} \right) \left(\frac{3030}{998} \right) = 17.9 \text{ J kg}^{-1} \tag{8.84}$$

The total friction loss is the sum of the component losses determined in the preceding calculations:

$$\sum F = 29.0 + 24.1 + 17.9 = 71.0 \text{ J kg}^{-1} \tag{8.85}$$

The work input to the pump can be determined by adding the total friction loss to the potential energy difference:

$$W = 9.81(3.5) + 71.0 = 105.3 \text{ J kg}^{-1} \tag{8.86}$$

The total system head is determined using this value of W:

$$H_s = \frac{W}{g} = \frac{105.3}{9.81} = 10.7 \text{ m} \tag{8.87}$$

5) Pressure at the pump entrance and the *NPSH* available. The friction losses prior to the pump must be calculated:

$$\sum F = \left[\frac{\bar{u}^2}{2}\left(k_{f,\text{ entrance}} + 3k_{f,\text{ elbow}}\right)\right]_{3\text{ inch line}} + \left[\frac{2f\bar{u}^2 L}{D}\right]_{3\text{ inch line}}$$

$$= \left[\frac{(1.70)^2}{2}\left(0.57 + 3(0.70)\right)\right] + \left[\frac{2(0.00733)(1.70)^2\, 2.5}{0.0720}\right] = 5.33 \text{ J kg}^{-1} \tag{8.88}$$

Using absolute pressure, P_1 = 1 atm = 101,420 Pa, and (from Eq.(8.68)) α = 1.12, the pressure at the pump inlet may be determined:

$$\left(P_2\right)_{\text{pump entrance}} = \rho\left(\frac{P_1}{\rho} - \frac{\bar{u}_2^2}{\alpha} - \sum F\right)$$

$$= 1030\left(\frac{101420}{1030} - \frac{(1.70)^2}{1.12} - 5.33\right) = 93.3 \text{ kPa (13.5 psia)} \tag{8.89}$$

The *NPSH* available is calculated using the pump inlet pressure found in Eq.(8.89), and the vapor pressure of water at the processing temperature (from Appendix 9.5, P_v = 1228 Pa at 10°C):

$$\left(NPSH\right)_A = \frac{\left(P_2\right)_{\text{pump entrance}} - P_v}{g\rho} = \frac{93{,}300 - 1228}{9.81(1030)} = 9.1 \text{ m} \tag{8.90}$$

6) Pressure at the pump exit.

$$P_{\text{pump exit}} = W\rho + P_{\text{pump entrance}}$$
$$= 105.3(1030) + 93300 = 201.8 \text{ kPa (29.3 psia)} \tag{8.91}$$

7) Total pressure drop across the pump.

$$\left(\Delta P\right)_{\text{over pump}} = W\rho$$
$$= 105.3(1030) = 108.5 \text{ kPa (15.7 psi)} \tag{8.92}$$

8) Hydraulic power requirement.

$$\Phi = W\dot{m} = 105.3(7.15) = 753 \text{ W} \quad (1 \text{ hp}) \tag{8.93}$$

8.9 Pumping a Shear Sensitive Fluid (Cream)

Cold (5°C) cream, a shear-sensitive fluid, is being pumped at 50 gpm in a straight section of 3 inch tubing (D = 2.83 inch = 0.072 m) using a diaphragm pump that introduces minimum shear degradation to the fluid; hence, only the mechanical degradation due to shear occurring in the pipeline must be evaluated. The following properties and parameters apply:

$\mu = 45$ cP = 0.045 Pa s; $\rho = 985$ kg m^{-3}

$Q = 50$ gpm = 0.00315 m^3 s^{-1}

$\bar{u} = 0.774$ m s^{-1} ; $\dot{m} = 3.10$ kg s^{-1}

$N_{\text{Re}} = 1219$; since flow is laminar, $f = 16/N_{\text{Re}}$

$L = 20$ m

1) Evaluate the shear power intensity in the present system.

2) Evaluate the shear power intensity in a new processing system where the length is tripled (L = 60 m) and volumetric flow rate remains constant at 50 gpm (Q = 0.00315 m^3 s^{-1}).

3) Evaluate the shear power intensity in a new processing system where the volumetric flow rate is increased by 50% (Q = 75 gpm = 0.00473 m^3 s^{-1}) and the length remains constant (L = 20 m). Also, ***assuming the original process is acceptable***, what new pipe diameter can be selected to increase the flow rate and retain an acceptable level of shear intensity for this system?

Solution

1) First evaluate the total shear work input (Eq. 6.1), then the shear power intensity (Eq. 6.2):

$$W_s = \sum \frac{\Phi}{\dot{m}} + \sum \frac{k_f \bar{u}^2}{2} + \sum \frac{2 f \bar{u}^2 L}{D} + \sum \left(\frac{\Delta P}{\rho} \right) \tag{8.94}$$

$$S = \frac{W_s \dot{m}}{V} \tag{8.95}$$

To simplify the discussion of these concepts, assume the only factor contributing significantly to W_s is the friction loss in the straight tubing. Then, Eq. (8.94) can be simplified to

$$W_s = \frac{2 f \bar{u}^2 L}{D} = \frac{2(16/1219)(0.774)^2(20)}{0.072} = 4.37 \text{ J kg}^{-1} \tag{8.96}$$

Using this value for shear work, the shear power intensity found in the 3 inch tubing can be evaluated:

$$S = \frac{W_s \dot{m}}{V} = \frac{4 W_s \dot{m}}{\pi D^2 L} = \frac{4(4.37)(3.10)}{\pi (0.072)^2 (20)} = 166 \text{ W m}^{-3} \tag{8.97}$$

where V is the volumetric capacity of a 20 m length of 3 inch tubing. Assume pumping under the above conditions (considered the "existing system" in subsequent discussion) produces an acceptable product.

2) In the case where the length is tripled (L = 60 m) and volumetric flow rate remains constant (Q = 50 gpm = 0.00315 m^3 s^{-1}), W_s and S for the tubing are

$$W_s = \frac{2f\bar{u}^2L}{D} = \frac{2(16/1219)(0.774)^2(60)}{0.072} = 13.1 \text{ J kg}^{-1} \quad (8.98)$$

$$S = \frac{W_s\dot{m}}{V} = \frac{4W_s\dot{m}}{\pi D^2 L} = \frac{4(13.1)(3.10)}{\pi(0.072)^2(60)} = 166 \text{ W m}^{-3} \quad (8.99)$$

Tripling the length caused W_s to triple but left S unchanged. Since the shear damage in cream is due to a disruption of the fat globule membrane, increasing W_s while leaving S unchanged has no adverse effect on the cream because the threshold rate of energy input per unit volume (established for the original system) is not exceeded.

3) In the case where the volumetric flow rate is increased by 50% (Q = 75 gpm = 0.00473 m^3 s^{-1}; $\dot{m}$ = 4.66 kg s^{-1}) and the length remains constant (L = 20 m), the mean velocity and Reynolds number are

$$\bar{u} = \frac{Q}{A} = \frac{0.00473(4)}{\pi(0.072)^2} = 1.16 \text{ m s}^{-1} \quad (8.100)$$

$$N_{\text{Re}} = \frac{\rho D\bar{u}}{\mu} = \frac{985(0.072)(1.16)}{0.045} = 1828 \quad (8.101)$$

With this information, W_s and S may be calculated at the increased flow:

$$W_s = \frac{2f\bar{u}^2L}{D} = \frac{2(16/1828)(1.16)^2(20)}{0.072} = 6.54 \text{ J kg}^{-1} \quad (8.102)$$

$$S = \frac{W_s\dot{m}}{V} = \frac{4W_s\dot{m}}{\pi D^2 L} = \frac{4(6.54)(4.66)}{\pi(0.072)^2(20)} = 374 \text{ W m}^{-3} \quad (8.103)$$

In this case, W_s is increased by 50% and S increased to 225% of the original value of 166 W m^{-3}. This high level of S is unacceptable because it may damage the product. To maintain the desired flow rate of

75 gpm, and an acceptable level of shear power intensity (166 W m^{-3} or less), the pipeline diameter must be increased. The value of C, and the diameter of the new pipe, can be calculated using the two flow rates under study and the scaling criterion established by Eq. (6.8):

$$C = \frac{Q_{\text{case 2}}}{Q_{\text{case 1}}} = \frac{75}{50} = 1.5 \tag{8.104}$$

$$D_{\text{case 2}} = C^{2/5} D_{\text{case 1}} = (1.5)^{2/5}(72) = 84.7 \text{ mm} \tag{8.105}$$

Since 84.7 mm internal diameter tubing is unavailable (see Appendix 9.3), the next larger size (an increase to 4 inch tubing) having an internal diameter of 97.4 mm (3.83 inches) is needed. Using this diameter results in new values for $\bar{u}$, N_{Re}, W_s, and S:

$$\bar{u} = \frac{Q}{A} = \frac{0.00473(4)}{\pi(0.0974)^2} = 0.635 \text{ m s}^{-1} \tag{8.106}$$

$$N_{\text{Re}} = \frac{\rho D \bar{u}}{\mu} = \frac{985(0.0974)(0.635)}{0.045} = 1354 \tag{8.107}$$

$$W_s = \frac{2 f \bar{u}^2 L}{D} = \frac{2(16/1354)(0.635)^2(20)}{0.0974} = 1.96 \text{ J kg}^{-1} \tag{8.108}$$

$$S = \frac{W_s \dot{m}}{V} = \frac{4 W_s \dot{m}}{\pi D^2 L} = \frac{4(1.96)(4.66)}{\pi(0.0974)^2(20)} = 61.3 \text{ W m}^{-3} \tag{8.109}$$

where V is the volumetric capacity of a 20 m section of 4 inch tubing. This value of S is significantly below the maximum allowable value of 166 W m^{-3}. In fact, with the low value of S found in the 4 inch pipe, an additional increase in the flow rate is possible without damaging the product. This result can be estimated if the equation for S is expanded by including the definition of W_s:

$$S = \frac{W_s \dot{m}}{V} = \frac{2 f \bar{u}^2 L}{D} \frac{\dot{m}}{V} \tag{8.110}$$

To find the velocity, assume laminar flow (so $f = 16/N_{\mathrm{Re}}$), substitute $\dot{m} = \bar{u} A \rho$ and $V = AL$, and solve for $\bar{u}$:

$$\bar{u} = \left(\frac{SD^2}{32\mu} \right)^{1/2} \tag{8.111}$$

Substituting known values gives the average velocity as

$$\bar{u} = \left(\frac{SD^2}{32\mu} \right)^{1/2} = \left(\frac{166(0.0974)^2}{32(0.045)} \right)^{1/2} = 1.05 \text{ m s}^{-1} \tag{8.112}$$

This velocity corresponds to a volumetric flow rate of 0.0078 $\text{m}^3 \text{ s}^{-1}$ (124 gpm). Checking the Reynolds number gives:

$$N_{\mathrm{Re}} = \frac{\rho D \bar{u}}{\mu} = \frac{985(0.0974)(1.05)}{0.045} = 2238 \tag{8.113}$$

This value of N_{Re} violates the assumption of laminar flow, so the Fanning friction factor must be determined for the transition regime: Using, Eq. (4.6), the Churchill equation $f = 0.00717$. Now, W_s and S may be recalculated:

$$W_s = \frac{2 f \bar{u}^2 L}{D} = \frac{2(0.00717)(1.05)^2(20)}{0.0974} = 3.25 \text{ J kg}^{-1} \tag{8.114}$$

$$S = \frac{W_s \dot{m}}{V} = \frac{4 W_s \dot{m}}{\pi D^2 L} = \frac{4(3.25)(7.68)}{\pi (0.0974)^2 (20)} = 167 \text{ W m}^{-3} \tag{8.115}$$

Using the 4 inch tube, a flow rate of 124 gpm is unlikely to damage the cream because the shear power intensity is held to an acceptable level; approximately 166 W m^{-3} as established by Eq. (8.99). It is interesting to note that the N_{Re} increased from 1354 to 2238 while S remained constant, clearly showing the Reynolds number was not constant during scale-up. Also, note that Example Problem 8.9 is presented to illustrate some of the technical issues involved in pumping shear-sensitive fluids: It is not intended to establish engineering design guidelines for cream handling systems.

8.10 Shear Power Intensity in a Centrifugal Pump

Example Problem 8.7 considered pumping cream from a truck to a storage tank. Assume the pump for this application is a centrifugal pump. Determine the pressure drop across the pump, and the shear power intensity delivered to the fluid by the pump, when the volumetric flow rate through the system is 50 gpm ($Q = 0.00315 \text{ m}^3 \text{ s}^{-1}$).

In solving Example Problem 8.7, the work input of the pump was 59.79 J kg^{-1} at a flow rate of 50 gpm. This information, and the density of cream ($\rho = 985 \text{ kg m}^{-3}$), can be used to calculate the pressure drop across the pump:

$$\Delta P = W\rho = 59.79(985) = 58.9 \text{ kPa} \ \ (8.54 \text{ psi}) \qquad (8.116)$$

The shear work (W_s) done on the fluid as it passes through the pump is equal to the work input of the pump: 59.79 J kg^{-1}. Knowing the shear work input, the shear power intensity generated in the pump can be calculated from Eq. 6.2:

$$S = \frac{W_s \dot{m}}{V} = \frac{59.79(0.00315)(985)}{0.0012} = 154.6 \, k\text{W m}^{-3} \qquad (8.117)$$

where V is the volumetric capacity of the pump under static conditions measured as 1.2 liters or 0.0012 m^3.

The total shear power intensity for the system includes the other components that generate a viscous dissipation of energy: entrance, exit, pneumatic valves, strainer, pipes, tee, and elbows. Eq. (8.59) indicates that these components of the system generate an additional shear work input of 25.45 J kg^{-1}.

8.11 Lethality of Pasteurization Process

A pilot scale system is pasteurizing milk at 10 gpm (1.34 ft^3 min^{-1}). In the heating section of the system, hot milk (ρ = 1110.2 kg m^{-3} = 63.05 lbm ft^{-3}; μ = 0.000400 Pa s = 0.000269 lbm ft^{-1} s^{-1}) exits a plate heat exchanger at 166°F, travels through a hold tube, then returns to the heat exchanger where it reenters the cooling section at 163°F. This part of the system is made of 2 inch stainless steel tubing (inside diameter = 1.87 inch) with the following lengths for each section: 3.17 ft from the heat exchanger to the hold tube, 27.25 ft of hold tube, and 3.92 ft from the hold tube back to the heat exchanger. Estimate the lethality of the process.

To solve this problem, the mean velocity (Eq. 4.2), the Reynolds number (Eq. 4.1), and the maximum velocity (Eq. 4.29) must be calculated:

$$\bar{u} = \frac{Q}{A} = \frac{1.34/60}{\dfrac{\pi}{4}\left(\dfrac{1.87}{12}\right)^2} = 1.17 \text{ ft s}^{-1} \tag{8.118}$$

$$N_{\mathrm{Re}} = \frac{D\bar{u}\rho}{\mu} = \frac{(1.87/12)1.17(63.05)}{0.000269} = 42735 \qquad (8.119)$$

$$u_{\max} = \frac{\bar{u}}{0.0336\log_{10}(N_{\mathrm{Re}}) + 0.662} = \frac{1.17}{0.0336\log_{10}(42735) + 0.662} = 1.43 \text{ ft s}^{-1} \qquad (8.120)$$

The analysis of the thermal process is based on the maximum velocity that, theoretically, represents the fastest moving milk particle traveling through the center of the pipe. Slower moving particles receive a greater heat treatment.

Assume the temperature drop is linear during the process. The total time the milk is subjected to high temperature is equal to the total length divided by the maximum velocity: $t_{process}$ = (3.17+27.25+3.92) / 1.43 = 24 s. The lethal rate is a function of temperature:

$$LR = 10^{\frac{T-161}{10.8}} \qquad (8.121)$$

assuming z = 10.8°F. Lethal rates, and lethalities, are summarized in Table 8.5.

The F value of the process is calculated using Eq. (7.24) and the result given in Table 8.5: $F_{161^\circ F}^{10.8^\circ F} = 50.2$ s. The legal requirement for the pasteurization of milk specifies a minimum heating time of 15s at 161°F. Using the above information, the process lethality can be calculated (Eq. 7.25):

$$\text{Process Lethality} = \frac{\left(F_{T_{ref}}^{z}\right)_{process}}{\left(F_{T_{ref}}^{z}\right)_{required}} = \frac{\left(F_{161^\circ F}^{10.8^\circ F}\right)_{process}}{\left(F_{161^\circ F}^{10.8^\circ F}\right)_{required}} = \frac{50.2}{15} = 3.35 \qquad (8.122)$$

meaning the process is 3.35 times more lethal than needed to eliminate pathogens. Although overheating may be undesirable for milk, it is not unusual since thermal processes may target spoilage organisms which may be more heat resistant than pathogens, or the material may be given additional heating to add a desirable cooked flavor to the product.

Table 8.5. Product temperatures and lethal rates over 2 s time intervals during the heating process.

t, s	*T*, °F	*LR*, s/s *	s at 161°F
2	165.8	2.8	5.5
4	165.5	2.6	5.2
6	165.3	2.5	5.0
8	165.0	2.3	4.7
10	164.8	2.2	4.5
12	164.5	2.1	4.2
14	164.3	2.0	4.0
16	164.0	1.9	3.8
18	163.8	1.8	3.6
20	163.5	1.7	3.4
22	163.3	1.6	3.2
24	163.0	1.5	3.1
		Total	50.2

* LR, s at 161°F / s at *T* °F

9 Appendices

9.1 Conversion Factors and Greek Alphabet

Density

1 g cm^{-3} = 1000 kg m^{-3} = 62.428 lbm ft^{-3} = 0.0361 lbm in^{-3}
1 lbm ft^{-3} = 16.0185 kg m^{-3}

Mass and Force

1 lbm = 16 oz = 0.45359 kg = 453.593 g
1 kg = 1000g = 0.001 metric ton = 2.20462 lbm = 35.274 oz
1 N = 1 kg m s^{-2} = 10^5 dyne = 10^5 g cm s^{-2} = 0.22481 lbf
1 lbf = 4.448 N = 32.174 lbm ft s^{-2}

Length

1 m = 100 cm = 1000 mm = 10^6 μm = 3.2808 ft = 39.37 in = 1.0936 yd
1 in = 2.54 cm = 25.40 mm = 0.0254 m = 0.0833 ft = 0.02778 yd

Power, Torque, and Energy

1 Btu = 1055 J = 1.055 kJ = 252.16 cal
1 hp = 550 ft lbf s^{-1} = 745.70 W = 0.7457 kW = 0.7068 Btu s^{-1}
1 W = 1 J s^{-1} = 0.23901 cal s^{-1} = 3.414 Btu h^{-1} = 1.341 (10^{-3}) hp
1 Btu hr^{-1} = 0.2931 W = 0.2931 J s^{-1}
1 N m = 1 kg m^2 s^{-2} = 10^7 dyne cm = 0.7376 ft lbf = 9.486 (10^{-4}) Btu
1 ft lbf = 1.35582 N m = 1.35582 J = 1.2851 (10^{-3}) Btu

Pressure and Stress

1 bar = 10^5 Pa = 14.5038 lbf in^{-2} = 0.987 atm = 33.48 ft H_2O at 4°C
1 Pa = 1 N m^{-2} = 10 dyne cm^{-2} = 9.8692 (10^{-6}) atm = 7.5 (10^{-3}) torr
1 lbf in^{-2} = 6894.8 Pa = 6.804 (10^{-2}) atm = 6.895 kPa
1 lbf in^{-2} = 2.309 ft H_2O = 2.0360 in. Hg
1 dyne cm^{-2} = 0.10 Pa = 10^{-6} bar = 0.987 (10^{-6}) atm
1 atm = 101.325 kPa = 14.696 psi = 1.013 bar = 29.921 in Hg @ 0°C
1 atm = 760 mm Hg at 0°C = 33.90 ft H_2O at 4°C

Temperature

$T_{Kelvin} = T_{Celsius} + 273.15$
$T_{Kelvin} = (T_{Fahrenheit} + 459.67) / 1.8$
$T_{Fahrenheit} = 1.8\ T_{Celsius} + 32$
$T_{Celsius} = (T_{Fahrenheit} - 32) / 1.8$

Viscosity (Absolute or Dynamic, followed by Kinematic)

1 P = 1 dyne s cm^{-2} = 0.1 Pa s = 100 cP = 100 mPa s
1 Pa s = 1000 cP = 10 P = 1 kg m^{-1} s^{-1} = 1 N s m^{-2} = 0.67197 lbm ft^{-1} s^{-1}
1 cP = 1 mPa s = 0.001 Pa s = 0.01 P
1 lbm ft^{-1} s^{-1} = 1.4882 kg m^{-1} s^{-1} = 1488.2 cP

kinematic viscosity: (cSt) = absolute viscosity (cP) / density (g cm^{-3})
1 cSt = 0.000001 $m^2\ s^{-1}$ = 1 $mm^2\ s^{-1}$ = 5.58001 $in^2\ hr^{-1}$ = 0.00155 $in^2\ s^{-1}$
1 St = 100 cSt = 0.0001 $m^2\ s^{-1}$
1 $m^2\ s^{-1}$ = 10^{-5} cSt = 10.7639 $ft^2\ s^{-1}$

Volume or Flow Rate
1 m^3 = 10^6 cm^3 = 10^3 L (liter) = 264.17 gal (US) = 35.3145 ft^3
1 ft^3 = 0.028317 m^3 = 7.4805 gal (US) = 28.317 L = 6.2288 gal(UK)
1 gal (US) = 128 oz (fluid) = 3.7854 L = 0.8327 gal (UK) = 0.003785 m^3
1 $m^3\ s^{-1}$ = 15,850.2 US gal min^{-1} = 264.17 US gal s^{-1}
1 US gal min^{-1} = 6.30902 (10^{-5}) $m^3\ s^{-1}$ = 3.7854 L min^{-1}
1 lbm hr^{-1} = 0.453 592 kg hr^{-1} = 1.25 998 (10^{-4}) kg s^{-1}

Greek Alphabet (upper and lower case)*

Α	α	alpha		Ν	ν	nu
Β	β	beta		Ξ	ξ	xi
Γ	γ	gamma		Ο	ο	omicron
Δ	δ	delta		Π	π	pi
δ	ε	epsilon		Ρ	ρ	rho
Ζ	ζ	zeta		Σ	σ	sigma
Η	η	eta		Τ	τ	tau
Θ	θ	theta		Υ	υ	upsilon
Ι	ι	iota		Φ	ϕ	phi
Κ	κ	kappa		Χ	χ	chi
Λ	λ	lambda		Ψ	ψ	psi
Μ	μ	mu		Ω	ω	Omega

* Dedicated to everyone (like the authors) who once referred to ξ or ζ as squiggle.

9.2 Rheological Properties of Biological Fluids

Product [#]	***T*, °C**	***K*, Pa s^n**	***n*** *
cream (40% fat)	40	0.0069	1.0
cream (40% fat)	60	0.0051	1.0
milk (raw)	5	0.00305	1.0
milk (raw)	25	0.00170	1.0
milk (raw)	40	0.00123	1.0
honey	6.5	76.6	1.0
honey	21.5	7.20	1.0
honey	48	0.50	1.0
olive oil	10	0.1380	1.0
olive oil	40	0.0363	1.0
olive oil	70	0.0124	1.0
soybean oil	20	0.0636	1.0
water	0	0.001787	1.0
water	25	0.000890	1.0
water	50	0.000547	1.0
water	75	0.000378	1.0
water	100	0.000281	1.0
whole egg	30	0.0064	1.0
apple sauce	30	11.6	0.34
Baby food puree	25	18.0	0.31
blueberry pie filling	20	6.08	0.43
caramel ice cream topping	40	35.6	0.57
chocolate	40	26.2	0.62
chocolate syrup	22	6.83	0.51
creamed corn	23	23.8	0.35
creamed corn	80	20.4	0.26
baby oil (creamy)	25	0.03	0.95
baby oil (creamy)	35	0.01	0.99
hair conditioner	25	16.1	0.19
hair conditioner	35	13.0	0.19
hair conditioner	45	11.5	0.14
hair styling gel	25	58.7	0.22
hand soap (liquid)	35	5.07	0.64
hand soap (liquid)	45	1.24	0.71

Product	T, °C	K, Pa s^n	n*
fudge ice cream topping	40	78.0	0.43
ice cream mix	1	0.34	0.76
ice cream mix	80	0.11	0.70
moisturizing cream	25	18.6	0.33
moisturizing cream	35	9.61	0.37
moisturizing cream	45	7.85	0.35
mustard	25	4.2	0.60
orange juice	30	0.14	0.79
orange juice concentrate	-18.8	24.4	0.76
orange juice concentrate	-5.4	6.45	0.77
orange juice concentrate	9.5	2.25	0.76
orange juice concentrate	29.2	0.69	0.80
pain relief ointment	25	7.31	0.47
pasta sauce (tomato)	23	40.3	0.19
pasta sauce (tomato)	80	21.8	0.18
pizza sauce	25	61.7	0.18
process cheese	25	320	0.50
process cheese	82	240	0.30
salsa (chunky)	23	30.8	0.20
salsa (chunky)	80	10.9	0.28
stomach reliever (bismuth subsalicylate)	25	0.43	0.81
sunscreen (creamy)	25	32.6	0.28
sunscreen (creamy)	35	24.3	0.29
sunscreen (creamy)	45	19.0	0.31
tomato ketchup	25	6.1	0.41

[#] Properties determined at typical pumping shear rates.

* Newtonian fluid if $n = 1.0$

9.3 Stainless Steel Tubing and Pipe Diameters

Nominal Size	Sanitary Tubing		Schedule 40 Pipe	
inch	I.D.* inch	O.D.# inch	I.D.* inch	O.D.# inch
1/2	0.37	0.50	0.62	0.84
3/4	0.62	0.75	0.82	1.05
1	0.87	1.00	1.05	1.32
1 1/2	1.37	1.50	1.61	1.90
2	1.87	2.00	2.07	2.38
2 1/2	2.37	2.50	2.47	2.88
3	2.83	3.00	3.07	3.50
4	3.83	4.00	4.03	4.50
6	5.78	6.00	6.07	6.63
8	7.78	8.00	7.98	8.63

inch	I.D.* mm	O.D.# mm	I.D.* mm	O.D.# mm
1/2	9.4	12.7	15.8	21.3
3/4	15.7	19.1	20.9	26.7
1	22.1	25.4	26.6	33.4
1 1/2	34.8	38.1	40.9	48.3
2	47.5	50.8	52.5	60.3
2 1/2	60.2	63.5	62.7	73.0
3	72.0	76.2	77.9	88.9
4	97.4	101.6	102.3	114.3
6	146.9	152.4	154.1	168.3
8	197.7	203.2	202.7	219.1

* I.D. = Inside Diameter

\# O.D. = Outside Diameter

9.4 Stainless Steel (304 and 316)

Steel is any alloy of iron and carbon that contains less than 2% carbon. Numbers 304 and 316 stainless steels are commonly used in food and pharmaceutical pipelines. They are non-magnetic materials (austenitic stainless steel) with a basic composition of 18% chromium and 8% nickel: the term "18/8 stainless" refers to this composition. Adding more chromium and nickel increases corrosion resistance. Small amounts of molybdenum greatly increase corrosion resistance. Numbers 304L and 316L are low carbon alloys developed to minimize carbide precipitation during welding which may lead to corrosion problems. Types 304 and 306 are designations of the American Iron and Steel Institute (AISI). Typical physical and thermal properties of 304 and 316 stainless steels include the following: density = 8,000 kg m^{-3}; thermal conductivity (at 100°C) = 16.3 W m^{-1} °C^{-1}; specific heat (0°C to 100°C) = 0.46 kJ kg^{-1}°C^{-1}; coefficient of thermal expansion (20°C to 500°C) = 17 x 10^{-6} m/m per °C.

Butt joints are a way of connecting two members lying in the same plane. Stainless steel tubing (including fittings, etc) may be joined with butt welds using a process known as gas tungsten-arc welding (GTAC). It is also called TIG (tungsten inert gas) welding. Heating is caused by an electric arc between the surface of the steel and a tungsten electrode. The electrode is not consumed during the process, and filling metal may be needed depending on the weld. An inert gas (such as argon) is used as a shielding gas to prevent atmospheric contamination of the weld. TIG welding requires a high degree of skill. It can produce high quality welds on many materials in addition to stainless steel: aluminum, magnesium, carbon steel, copper, nickel, and titanium.

Chemical composition of typical stainless steel (Source: ASTM A270)

Element	304	304L	316	316L
	*Composition, %			
Carbon, max	0.08	0.035	0.08	0.035
Manganese, max	2.00	2.00	2.00	2.00
Phosphorus, max	0.040	0.040	0.040	0.040
Sulfur, max	0.030	0.030	0.030	0.030
Silcon, max	0.75	0.75	0.030	0.030
Nickel	8.00 - 11.00	8.00 - 13.00	10.00 - 14.00	10.00 - 15.00
Chromium	18.00 - 20.00	18.00 - 20.00	16.00 - 18.00	16.00 - 18.00
Molybdenum	none	none	2.00 - 3.00	2.00 - 3.00

* Balance of composition is iron.

9.5 Properties of Saturated Water

T °C	T °F	abs P* kPa	abs P* psi	vacuum in Hg	density kg/m^3	density lbm/ft^3
0.01	32.018	0.6117	0.089	29.74	999.8	62.42
5	41	0.8726	0.127	29.66	999.9	62.42
10	50	1.228	0.178	29.56	999.7	62.41
15	59	1.706	0.247	29.42	999.1	62.37
20	68	2.339	0.339	29.23	998.2	62.32
25	77	3.170	0.460	28.98	997.0	62.24
30	86	4.247	0.616	28.67	995.6	62.15
35	95	5.629	0.816	28.26	994.0	62.06
40	104	7.384	1.071	27.74	992.2	61.94
45	113	9.594	1.391	27.09	990.2	61.82
50	122	12.351	1.791	26.27	988.0	61.68
55	131	15.761	2.286	25.27	985.7	61.54
60	140	19.946	2.893	24.03	983.2	61.38
65	149	25.041	3.632	22.53	980.5	61.21
70	158	31.201	4.525	20.71	977.7	61.04
75	167	38.595	5.598	18.52	974.8	60.86
80	176	47.415	6.877	15.92	971.8	60.67
85	185	57.867	8.393	12.83	968.6	60.47
90	194	70.182	10.179	9.20	965.3	60.26
95	203	84.609	12.271	4.94	961.9	60.05
100	212	101.42	14.710	0	958.3	59.83
105	221	120.90	17.535	0	954.7	59.60
110	230	143.38	20.795	0	950.9	59.36
115	239	169.18	24.537	0	947.1	59.12
120	248	198.67	28.815	0	943.1	58.88

T °C	T °F	abs P* kPa	abs P* psi	vacuum in Hg	density kg/m^3	density lbm/ft^3
125	257	232.22	33.681	0	939.1	58.62
130	266	270.26	39.198	0	934.8	58.36
135	275	313.20	45.426	0	930.5	58.09
140	284	361.50	52.431	0	926.1	57.81
145	293	415.63	60.282	0	921.7	57.54
150	302	476.10	69.052	0	917.0	57.25
155	311	543.57	78.838	0	912.2	56.95
160	320	618.14	89.653	0	907.4	56.65
165	329	700.99	101.67	0	902.5	56.34
170	338	792.05	114.88	0	897.4	56.02
175	347	892.65	129.47	0	892.3	55.70
180	356	1000.3	145.08	0	887.0	55.37
185	365	1123.5	162.95	0	881.6	55.04
190	374	1255.0	182.02	0	876.1	54.69
195	383	1398.9	202.89	0	870.4	54.34
200	392	1554.7	225.49	0	864.7	53.98
205	401	1724.4	250.10	0	858.7	53.61

* absolute pressure

Source: ASME, 2000. ASME International Steam Tables for Industrial Use. The American Society of Mechanical Engineers, ASME Press, New York.

9.6 Enthalpy of Saturated Steam

T	T	abs P*	abs P*	Enthalpy (kJ/kg)		
°C	°F	kPa	psi	h_L	Δh	h_v
0.01	32.018	0.6117	0.09	0.001	2500.9	2500.9
5	41	0.8726	0.13	21.019	2489.1	2510.1
10	50	1.228	0.18	42.021	2477.2	2519.2
15	59	1.706	0.25	62.984	2465.4	2528.4
20	68	2.339	0.34	83.92	2453.5	2537.5
25	77	3.170	0.46	104.84	2441.7	2546.5
30	86	4.247	0.62	125.75	2429.8	2555.6
35	95	5.629	0.82	146.64	2417.9	2564.6
40	104	7.384	1.07	167.54	2406.0	2573.5
45	113	9.594	1.39	188.44	2394.0	2582.5
50	122	12.351	1.79	209.34	2382.0	2591.3
55	131	15.761	2.29	230.24	2369.9	2600.0
60	140	19.946	2.89	251.15	2357.7	2608.8
65	149	25.041	3.63	272.08	2345.4	2617.5
70	158	31.201	4.53	293.02	2333.1	2626.1
75	167	38.595	5.60	313.97	2320.6	2634.6
80	176	47.415	6.88	334.95	2308.1	2643.0
85	185	57.867	8.39	355.95	2295.4	2651.3
90	194	70.182	10.18	376.97	2282.6	2659.5
95	203	84.609	12.27	398.02	2269.6	2667.6
100	212	101.42	14.71	419.10	2256.5	2675.6
105	221	120.90	17.54	440.21	2243.2	2683.4
110	230	143.38	20.80	461.36	2229.7	2691.1
115	239	169.18	24.54	482.55	2216.0	2698.6
120	248	198.67	28.81	503.78	2202.1	2705.9
125	257	232.22	33.68	525.06	2188.0	2713.1

T	T	abs P*	abs P*	Enthalpy (kJ/kg)		
°C	°F	kPa	psi	h_L	Δh	h_v
130	266	270.26	39.20	546.39	2173.7	2720.1
135	275	313.20	45.43	567.77	2159.1	2726.9
140	284	361.50	52.43	589.20	2144.2	2733.4
145	293	415.63	60.28	610.69	2129.1	2739.8
150	302	476.10	69.05	632.25	2113.7	2745.9
155	311	543.57	78.84	653.88	2097.9	2751.8
160	320	618.14	89.65	675.57	2081.9	2757.4
165	329	700.99	101.67	697.35	2065.5	2762.8
170	338	792.05	114.88	719.21	2048.7	2767.9
175	347	892.65	129.47	741.15	2031.6	2772.7
180	356	1000.3	145.07	763.19	2014.0	2777.2
185	365	1123.5	162.95	785.33	1996.1	2781.4
190	374	1255.0	182.02	807.57	1977.7	2785.3
195	383	1398.9	202.89	829.92	1958.9	2788.9
200	392	1554.7	225.49	852.39	1939.7	2792.1
205	401	1724.4	250.09	874.88	1919.9	2794.9

* absolute pressure

Source: ASME, 2000. ASME International Steam Tables for Industrial Use. The American Society of Mechanical Engineers, ASME Press, New York.

9.7 Viscosity (mPa s or cP) of Water

T (°C)	Pressure (MPa)							
	0.01	**0.02**	**0.05**	**0.1**	**0.2**	**0.5**	**1**	**2**
0	1.792	1.792	1.792	1.792	1.791	1.791	1.789	1.787
10	1.306	1.306	1.306	1.306	1.306	1.305	1.305	1.304
20	1.002	1.002	1.002	1.002	1.002	1.001	1.001	1.009
25	0.890	0.890	0.890	0.890	0.890	0.890	0.890	0.890
30	0.797	0.797	0.797	0.797	0.797	0.797	0.797	0.797
40	0.653	0.653	0.653	0.653	0.653	0.653	0.653	0.653
50		0.547	0.547	0.547	0.547	0.547	0.547	0.547
60		0.466	0.466	0.466	0.466	0.467	0.467	0.467
70			0.404	0.404	0.404	0.404	0.404	0.404
80			0.354	0.354	0.354	0.355	0.355	0.355
90				0.314	0.314	0.315	0.315	0.315
100					0.282	0.282	0.282	0.282
110					0.255	0.255	0.255	0.255
120					0.232	0.232	0.232	0.233
130						0.213	0.213	0.213
140						0.197	0.197	0.197
150						0.183	0.183	0.183
160							0.170	0.171
170							0.160	0.160
180								0.150
190								0.142
200								0.134

Note: 1 atm = 0.101 MPa = 14.96 psi

Source: ASME, 2000. ASME International Steam Tables for Industrial Use. The American Society of Mechanical Engineers, ASME Press, New York.

9.8 Gallons of Water per 100 feet of Tubing.

Nominal Size, inch	I.D. inch	O.D. inch	gal per 100 ft of pipe
1/2	0.370	0.500	0.6
3/4	0.620	0.750	1.6
1	0.870	1.000	3.1
1 1/2	1.370	1.500	7.7
2	1.870	2.000	14.3
2 1/2	2.370	2.500	22.9
3	2.834	3.000	32.8
4	3.834	4.000	60.0
6	5.782	6.000	136.4
8	7.782	8.000	247.1

9.9 Affinity Laws for Centrifugal Pumps

For small variations in impeller diameter (at constant speed):

$$\frac{D_1}{D_2} = \frac{Q_1}{Q_2} = \frac{(H_1)^{1/2}}{(H_2)^{1/2}} \tag{9.1}$$

$$\frac{bhp_1}{bhp_2} = \frac{D_1^{\,3}}{D_2^{\,3}} \tag{9.2}$$

For small variations in speed (at constant impeller diameter):

$$\frac{N_1}{N_2} = \frac{Q_1}{Q_2} = \frac{(H_1)^{1/2}}{(H_2)^{1/2}} \tag{9.3}$$

$$\frac{bhp_1}{bhp_2} = \frac{N_1^{\,3}}{N_2^{\,3}} \tag{9.4}$$

where: D = impeller diameter; H = head; Q = volumetric capacity; N = angular velocity; bhp = break horsepower.

9.10 Equations for Bingham Plastic Fluids in Tube Flow

Fanning Friction Factors. The equation of state for a Bingham Plastic is

$$\sigma = \mu_{pl}\dot{\gamma} + \sigma_o \tag{9.5}$$

where μ_{pl} = plastic viscosity (Pa s); σ_o = yield stress (Pa). The presence of a yield stress makes this model unique.

Laminar Flow Criterion:

$$N_{\mathrm{Re},B} \le \left(N_{\mathrm{Re},B}\right)_{critical} \tag{9.6}$$

$$\left(N_{\mathrm{Re},B}\right)_{critical} = \frac{N_{He}}{8c_c}\left(1 - \frac{4c_c}{3} + \frac{c_c^{\,4}}{3}\right) \tag{9.7}$$

$$\frac{c_c}{\left(1-c_c\right)^3} = \frac{N_{He}}{16{,}800} \tag{9.8}$$

$$c = \frac{\sigma_o}{\sigma_w} = \frac{R_o}{R} \tag{9.9}$$

where the Hedstrom Number (N_{He}), the Bingham Reynolds Number ($N_{\mathrm{Re},B}$) and the shear stress at the wall (σ_w) are:

$$N_{He} = \frac{D^2 \sigma_o \rho}{\left(\mu_{pl}\right)^2} \tag{9.10}$$

$$N_{\mathrm{Re},B} = \frac{D\bar{u}\rho}{\mu_{pl}} \tag{9.11}$$

and

$$\sigma_w = \frac{\Delta P\, R}{2L} \tag{9.12}$$

The laminar flow friction factor [from Heywood, N. 1991. Pipeline design for non-settling slurries. In Brown, N.P. and N.I. Heywood (editors). Slurry Handling: Design of Solid-Liquid Systems. Elsevier Applied Science, NY] assuming $\left(\sigma_o / \sigma_w\right)^4 << 1$ is

$$f_{\text{laminar}} = \frac{16\left(6N_{\text{Re},B} + N_{He}\right)}{6\left(N_{\text{Re},B}\right)^2} \tag{9.13}$$

The turbulent flow friction factor [from Darby, R., Mun, R., Boger, V.B. 1992. Prediction friction loss in slurry pipes. *Chemical Engineering*, September, pg 116-119] is

$$f_{\text{turbulent}} = \frac{10^a}{\left(N_{\text{Re},B}\right)^{0.193}} \tag{9.14}$$

where:

$$a = -1.47\left[1 + 0.146\exp\left(-2.9x10^{-5} N_{He}\right)\right] \tag{9.15}$$

Velocity Profiles in Laminar Tube Flow. The velocity profile in the sheared portion ($R_o \leq r \leq R \quad where \quad \sigma \geq \sigma_o$) of the fluid in the pipe is

$$u = \frac{(\Delta P)R^2}{4\mu_{pl}L}\left[1 - \left(\frac{r}{R}\right)^2 - \frac{2R_o}{R}\left(1 - \frac{r}{R}\right)\right] \tag{9.16}$$

where the critical radius (R_o) is calculated in terms of the yield stress:

$$R_o = \frac{2L\sigma_o}{\Delta P} \tag{9.17}$$

The maximum velocity (u = u_{max}) is located in the unsheared plug ($0 \le r \le R_0$ where $\sigma \le \sigma_o$):

$$u_{max} = \frac{(\Delta P) R^2}{4\mu_{pl} L}\left(1 - \frac{R_o}{R}\right)^2 \tag{9.18}$$

or

$$u_{max} = \bar{u}\left(\frac{2(1-c)^2}{1 - \frac{4c}{3} + \frac{c^4}{3}}\right) \tag{9.19}$$

9.11 Fanning Friction Factors for Power Law Fluids

$N_{Re, PL}$	$n = 0.2$	$n = 0.3$	$n = 0.4$	$n = 0.5$	$n = 0.6$	$n = 0.7$	$n = 0.8$	$n = 0.9$
2100	0.00762	0.00762	0.00762	0.00762	0.00762	0.00762	0.00762	0.00762
2200	0.00727	0.00727	0.00727	0.00727	0.00727	0.00727	0.00727	0.00727
2300	0.00696	0.00696	0.00696	0.00696	0.00696	0.00696	0.00717	0.00763
2400	0.00667	0.00667	0.00667	0.00667	0.00667	0.00697	0.00745	0.00795
2500	0.00640	0.00640	0.00640	0.00640	0.00670	0.00719	0.00771	0.00825
2600	0.00615	0.00615	0.00615	0.00635	0.00686	0.00739	0.00794	0.00851
2700	0.00593	0.00593	0.00592	0.00645	0.00700	0.00756	0.00815	0.00875
2800	0.00571	0.00571	0.00596	0.00571	0.00711	0.00571	0.00832	0.00571
2900	0.00489	0.00541	0.00599	0.00659	0.00720	0.00782	0.00845	0.00909
3000	0.00487	0.00540	0.00601	0.00663	0.00726	0.00791	0.00855	0.00920
3100	0.00483	0.00538	0.00600	0.00665	0.00730	0.00796	0.00862	0.00927
3200	0.00480	0.00535	0.00599	0.00665	0.00732	0.00799	0.00866	0.00931
3300	0.00475	0.00532	0.00597	0.00664	0.00733	0.00800	0.00867	0.00933
3400	0.00471	0.00528	0.00594	0.00663	0.00731	0.00800	0.00867	0.00932
3500	0.00467	0.00524	0.00591	0.00660	0.00729	0.00798	0.00865	0.00930
3600	0.00462	0.00520	0.00587	0.00656	0.00726	0.00796	0.00863	0.00927
3700	0.00457	0.00515	0.00583	0.00653	0.00723	0.00792	0.00859	0.00924
3800	0.00453	0.00511	0.00578	0.00649	0.00719	0.00788	0.00855	0.00920
3900	0.00448	0.00506	0.00574	0.00645	0.00715	0.00784	0.00851	0.00915
4000	0.00444	0.00502	0.00570	0.00640	0.00711	0.00780	0.00847	0.00910
4100	0.00440	0.00498	0.00565	0.00636	0.00706	0.00775	0.00842	0.00906
4200	0.00435	0.00493	0.00561	0.00631	0.00702	0.00771	0.00837	0.00901
4300	0.00431	0.00489	0.00557	0.00627	0.00698	0.00766	0.00833	0.00896
4400	0.00427	0.00485	0.00553	0.00623	0.00693	0.00762	0.00828	0.00891
4500	0.00423	0.00481	0.00548	0.00619	0.00689	0.00757	0.00823	0.00886
4600	0.00420	0.00477	0.00544	0.00614	0.00684	0.00753	0.00819	0.00882
4700	0.00416	0.00473	0.00540	0.00610	0.00680	0.00749	0.00814	0.00877
4800	0.00412	0.00470	0.00537	0.00606	0.00676	0.00744	0.00810	0.00873
4900	0.00409	0.00466	0.00533	0.00602	0.00672	0.00740	0.00806	0.00868
5000	0.00405	0.00462	0.00529	0.00599	0.00668	0.00736	0.00802	0.00864
5100	0.00402	0.00459	0.00525	0.00595	0.00664	0.00732	0.00798	0.00860
5200	0.00398	0.00456	0.00522	0.00591	0.00660	0.00728	0.00793	0.00856
5300	0.00395	0.00452	0.00518	0.00587	0.00657	0.00724	0.00790	0.00852
5400	0.00392	0.00449	0.00515	0.00584	0.00653	0.00721	0.00786	0.00848

$N_{Re,PL}$	**n = 0.2**	**n = 0.3**	**n = 0.4**	**n = 0.5**	**n = 0.6**	**n = 0.7**	**n = 0.8**	**n = 0.9**
5500	0.00389	0.00446	0.00512	0.00580	0.00649	0.00717	0.00782	0.00844
5600	0.00386	0.00443	0.00508	0.00577	0.00646	0.00713	0.00778	0.00840
5700	0.00383	0.00440	0.00505	0.00574	0.00642	0.00710	0.00775	0.00837
5800	0.00380	0.00437	0.00502	0.00571	0.00639	0.00706	0.00771	0.00833
5900	0.00378	0.00434	0.00499	0.00567	0.00636	0.00703	0.00768	0.00830
6000	0.00375	0.00431	0.00496	0.00564	0.00633	0.00700	0.00764	0.00826
6100	0.00372	0.00428	0.00493	0.00561	0.00629	0.00696	0.00761	0.00823
6200	0.00370	0.00426	0.00491	0.00558	0.00626	0.00693	0.00758	0.00819
6300	0.00367	0.00423	0.00488	0.00555	0.00623	0.00690	0.00754	0.00816
6400	0.00365	0.00421	0.00485	0.00553	0.00620	0.00687	0.00751	0.00813
6500	0.00363	0.00418	0.00482	0.00550	0.00618	0.00684	0.00748	0.00810
6600	0.00360	0.00416	0.00480	0.00547	0.00615	0.00681	0.00745	0.00807
6700	0.00358	0.00413	0.00477	0.00544	0.00612	0.00678	0.00742	0.00804
6800	0.00356	0.00411	0.00475	0.00542	0.00609	0.00675	0.00739	0.00801
6900	0.00353	0.00409	0.00472	0.00539	0.00606	0.00673	0.00737	0.00798
7000	0.00351	0.00406	0.00470	0.00537	0.00604	0.00670	0.00734	0.00795
7100	0.00349	0.00404	0.00468	0.00534	0.00601	0.00667	0.00731	0.00792
7200	0.00347	0.00402	0.00465	0.00532	0.00599	0.00665	0.00728	0.00790
7300	0.00345	0.00400	0.00463	0.00529	0.00596	0.00662	0.00726	0.00787
7400	0.00343	0.00398	0.00461	0.00527	0.00594	0.00659	0.00723	0.00784
7500	0.00341	0.00396	0.00459	0.00525	0.00591	0.00657	0.00720	0.00782
7600	0.00339	0.00394	0.00457	0.00523	0.00589	0.00654	0.00718	0.00779
7700	0.00337	0.00392	0.00454	0.00520	0.00587	0.00652	0.00715	0.00776
7800	0.00335	0.00390	0.00452	0.00518	0.00584	0.00650	0.00713	0.00774
7900	0.00334	0.00388	0.00450	0.00516	0.00582	0.00647	0.00711	0.00772
8000	0.00332	0.00386	0.00448	0.00514	0.00580	0.00645	0.00708	0.00769
8100	0.00330	0.00384	0.00446	0.00512	0.00578	0.00643	0.00706	0.00767
8200	0.00328	0.00382	0.00444	0.00510	0.00576	0.00641	0.00704	0.00764
8300	0.00327	0.00380	0.00443	0.00508	0.00574	0.00638	0.00701	0.00762
8400	0.00325	0.00379	0.00441	0.00506	0.00571	0.00636	0.00699	0.00760
8500	0.00323	0.00377	0.00439	0.00504	0.00569	0.00634	0.00697	0.00758
8600	0.00322	0.00375	0.00437	0.00502	0.00567	0.00632	0.00695	0.00755
8700	0.00320	0.00374	0.00435	0.00500	0.00565	0.00630	0.00693	0.00753
8800	0.00319	0.00372	0.00433	0.00498	0.00563	0.00628	0.00691	0.00751
8900	0.00317	0.00370	0.00432	0.00496	0.00562	0.00626	0.00689	0.00749
9000	0.00316	0.00369	0.00430	0.00495	0.00560	0.00624	0.00687	0.00747
9100	0.00314	0.00367	0.00428	0.00493	0.00558	0.00622	0.00685	0.00745

$N_{Re, PL}$	**n = 0.2**	**n = 0.3**	**n = 0.4**	**n = 0.5**	**n = 0.6**	**n = 0.7**	**n = 0.8**	**n = 0.9**
9200	0.00313	0.00366	0.00427	0.00491	0.00556	0.00620	0.00683	0.00743
9300	0.00311	0.00364	0.00425	0.00489	0.00554	0.00618	0.00681	0.00741
9400	0.00310	0.00363	0.00423	0.00488	0.00552	0.00616	0.00679	0.00739
9500	0.00308	0.00361	0.00422	0.00486	0.00551	0.00615	0.00677	0.00737
9600	0.00307	0.00360	0.00420	0.00484	0.00549	0.00613	0.00675	0.00735
9700	0.00306	0.00358	0.00419	0.00483	0.00547	0.00611	0.00673	0.00733
9800	0.00304	0.00357	0.00417	0.00481	0.00545	0.00609	0.00671	0.00731
9900	0.00303	0.00355	0.00416	0.00479	0.00544	0.00607	0.00670	0.00729
10000	0.00302	0.00354	0.00414	0.00478	0.00542	0.00606	0.00668	0.00728
12000	0.00279	0.00330	0.00388	0.00450	0.00513	0.00575	0.00636	0.00695
14000	0.00261	0.00311	0.00368	0.00428	0.00490	0.00551	0.00611	0.00669
16000	0.00247	0.00295	0.00351	0.00410	0.00470	0.00530	0.00590	0.00647
18000	0.00235	0.00282	0.00337	0.00394	0.00454	0.00513	0.00572	0.00629
20000	0.00225	0.00271	0.00324	0.00381	0.00439	0.00498	0.00556	0.00612
22000	0.00216	0.00261	0.00313	0.00369	0.00427	0.00485	0.00542	0.00598
24000	0.00208	0.00252	0.00304	0.00359	0.00416	0.00473	0.00530	0.00585
26000	0.00201	0.00245	0.00295	0.00350	0.00406	0.00463	0.00519	0.00574
28000	0.00195	0.00238	0.00288	0.00341	0.00397	0.00453	0.00509	0.00563
30000	0.00189	0.00232	0.00281	0.00334	0.00389	0.00444	0.00499	0.00554
32000	0.00184	0.00226	0.00275	0.00327	0.00381	0.00436	0.00491	0.00545
34000	0.00179	0.00221	0.00269	0.00321	0.00374	0.00429	0.00483	0.00537
36000	0.00175	0.00216	0.00263	0.00315	0.00368	0.00422	0.00476	0.00529
38000	0.00171	0.00211	0.00258	0.00309	0.00362	0.00416	0.00469	0.00522
40000	0.00167	0.00207	0.00254	0.00304	0.00356	0.00410	0.00463	0.00515
42000	0.00164	0.00203	0.00249	0.00299	0.00351	0.00404	0.00457	0.00509
44000	0.00161	0.00200	0.00245	0.00295	0.00346	0.00399	0.00451	0.00503
46000	0.00158	0.00196	0.00241	0.00290	0.00342	0.00394	0.00446	0.00498
48000	0.00155	0.00193	0.00238	0.00286	0.00337	0.00389	0.00441	0.00493
50000	0.00152	0.00190	0.00234	0.00283	0.00333	0.00385	0.00436	0.00488
52000	0.00150	0.00187	0.00231	0.00279	0.00329	0.00380	0.00432	0.00483
54000	0.00147	0.00184	0.00228	0.00276	0.00325	0.00376	0.00428	0.00478
56000	0.00145	0.00182	0.00225	0.00272	0.00322	0.00372	0.00423	0.00474
58000	0.00143	0.00179	0.00222	0.00269	0.00318	0.00369	0.00420	0.00470
60000	0.00141	0.00177	0.00220	0.00266	0.00315	0.00365	0.00416	0.00466
62000	0.00139	0.00175	0.00217	0.00263	0.00312	0.00362	0.00412	0.00462
64000	0.00137	0.00173	0.00215	0.00261	0.00309	0.00359	0.00409	0.00459
66000	0.00135	0.00171	0.00212	0.00258	0.00306	0.00356	0.00405	0.00455

$N_{Re, PL}$	**n = 0.2**	**n = 0.3**	**n = 0.4**	**n = 0.5**	**n = 0.6**	**n = 0.7**	**n = 0.8**	**n = 0.9**
68000	0.00133	0.00169	0.00210	0.00256	0.00303	0.00353	0.00402	0.00452
70000	0.00132	0.00167	0.00208	0.00253	0.00301	0.00350	0.00399	0.00448
72000	0.00130	0.00165	0.00206	0.00251	0.00298	0.00347	0.00396	0.00445
74000	0.00129	0.00163	0.00204	0.00249	0.00296	0.00344	0.00393	0.00442
76000	0.00127	0.00162	0.00202	0.00247	0.00293	0.00342	0.00391	0.00439
78000	0.00126	0.00160	0.00200	0.00244	0.00291	0.00339	0.00388	0.00437
80000	0.00124	0.00158	0.00199	0.00242	0.00289	0.00337	0.00385	0.00434
82000	0.00123	0.00157	0.00197	0.00240	0.00287	0.00334	0.00383	0.00431
84000	0.00122	0.00156	0.00195	0.00239	0.00285	0.00332	0.00380	0.00429
86000	0.00121	0.00154	0.00193	0.00237	0.00283	0.00330	0.00378	0.00426
88000	0.00120	0.00153	0.00192	0.00235	0.00281	0.00328	0.00376	0.00424
90000	0.00118	0.00151	0.00190	0.00233	0.00279	0.00326	0.00373	0.00421
92000	0.00117	0.00150	0.00189	0.00232	0.00277	0.00324	0.00371	0.00419
94000	0.00116	0.00149	0.00187	0.00230	0.00275	0.00322	0.00369	0.00417
96000	0.00115	0.00148	0.00186	0.00228	0.00273	0.00320	0.00367	0.00415
98000	0.00114	0.00147	0.00185	0.00227	0.00272	0.00318	0.00365	0.00412

Based on Eq. (4.16) to (4.21) from Darby, R., Mun, R., Boger, V.B. 1992. Chemical Engineering, September, pg 116-119.

9.12 Friction Loss Coefficients: 3-k equation

The 3-k equation is

$$k_f = \frac{k_1}{N_{\text{Re}}} + k_2\left(1 + \frac{k_3}{\left(D_{n,in.}\right)^{0.3}}\right) \tag{9.20}$$

where: $D_{n,in.}$ = nominal diameter, inches, and k_3 has dimensions of (inches)$^{0.3}$. Use $N_{Re,PL}$ in place of N_{Re} for power law fluids. Note: this equation was formulated for standard steel pipe. Using nominal diameters for tubing will produce slightly higher values of the friction loss coefficient.

Valve or Fitting #		k_1	k_2	k_3
elbow 90°: threaded, standard	$r/D = 1$	800	0.14	4.0
elbow 90°: threaded, long radius	$r/D = 1.5$	800	0.071	4.2
elbow 90°: flanged, butt welded	$r/D = 1$	800	0.091	4.0
elbow 90°: flanged, butt welded	$r/D = 2$	800	0.056	3.9
elbow 90°: flanged, butt welded	$r/D = 4$	800	0.066	3.9
elbow 90°: flanged, butt welded	$r/D = 6$	800	0.0075	4.2
elbow 45°: threaded, standard	$r/D = 1$	500	0.071	4.2
elbow 45°: threaded, long radius	$r/D = 1.5$	500	0.052	4.0
elbow 180°: threaded, close return bend	$r/D = 1$	1000	0.23	4.0
elbow 180°: flanged, butt welded	$r/D = 1$	1000	0.12	4.0
elbow 180°: all	$r/D = 1.5$	1000	0.10	4.0
tee: threaded, through branch (as elbow)	$r/D = 1$	500	0.274	4.0
tee: threaded, through branch (as elbow)	$r/D = 1.5$	800	0.14	4.0
tee: flanged/welded, through branch (as elbow)	$r/D = 1$	800	0.28	4.0
tee: threaded, run through	$r/D = 1$	200	0.091	4.0
tee: flanged/welded, run through	$r/D = 1$	150	0.017	4.0

Valve or Fitting [#]		k_1	k_2	k_3
angle valve - 45°	β=1[*]	950	0.25	4.0
angle valve - 90°	β=1[*]	1000	0.69	4.0
globe valve, standard	β=1[*]	1500	1.70	3.6
plug valve, branch flow		500	0.41	4.0
plug valve, straight through		300	0.084	3.9
plug valve, three-way (flow through)		300	0.14	4.0
gate valve, standard	β=1[*]	300	0.037	3.9
ball valve, standard	β=1[*]	300	0.015	3.5
diaphragm, dam-type		1000	0.69	4.9

Source: Darby, R. 2001 (March). Chemical Engineering, 66-73

[#] Except for some high-pressure applications, threaded product contact surfaces are not allowed in sanitary systems (3A Standard Number 63-10). Standard threaded steel is acceptable for water supply systems.

[*] β = orifice diameter / pipe inside diameter.

Index

12*D* process, 92
3-A Sanitary Standards, 49
anti-thixotropic fluid, 7
apparent viscosity, 5
Arrhenius equation, 12, 96
average shear rate. *See* shear rate
average shear stress. *See* shear stress
Bernoulli equation, 35
Bingham plastic, 6, 149
Bingham Reynolds Number, 149
Blasius equation, 50
Bot Cook, 92
brake horsepower, 37
Brookfield viscometer, 30
Casson equation, 6, 99
centrifugal pump, 45, 132
Churchill equation, 117
Clostridium botulinum, 87, 91, 92
concentric cylinder viscometer, 15, 102
cone and plate viscometer, 33
Coxiellia burnettii, 87, 89, 93
D value, 86
Darcy friction factor, 49
death kinetics, 85
decimal reductions, 87
DIN, 15
effective viscosity, 110
elevation head, 38, 48
end effect correction, 18
equivalent length method, 37, 61
F value, 89
Fanning friction factor, 36, 49, 56, 57
friction head, 38
friction loss coefficient, 61
friction losses, 35
general method, 89, 95
Haaland equation, 117
head, 38, 39, 40, 41, 48, 125
Hedstrom Number, 149
Herschel-Bulkley equation, 6, 99
hydraulic diameter, 36
hydraulic power, 37, 119
impeller
- Brookfield disk, *31*
- Brookfield flag, 27
- Haake pitched paddle, 27
- helical ribbon, 25
- interrupted helical screw, 25, 107
- RVA pitched paddle, 27

impeller Reynolds number, 20
k'' method, 20
kinematic viscosity, 4
kinetic energy correction factor, 36
kinetic energy loss, 35
lethal rate, 91, 95, 134
lethality, 95, 133, 134
low acid food, 88
matching viscosity assumption, 22, 29
mechanical energy balance, 35, 115
microbial inactivation, 85
minimum shear rate, 96
mixer coefficient, 20, 21, 24, 106
mixer viscometer constant, 20, 22, 24
mixer viscometry, 19, 107
net positive suction head, 46, 122
Newtonian fluid, 3, 49, 96
parallel plate viscometer, 34
parallel plates, 4
particle size, 24
pasteurization, *89*, 90, *93*, 133
pipeline scale-up, 82, *84*
Poiseuille-Hagen equation, 9
positive displacement pump, 46, 120
potential energy loss, 35
power law fluid, *5*, 51, 100
power law impeller Reynolds number, 25

power law Reynolds number, 52, 110
power number, 20
pressure energy loss, 35
pressure head, 38, 48
process lethality
  see lethality, *95*
pump curve, 45
pump efficiency, 37
pump head, 45
relative kill time, 91
relative roughness, 51
representative shear rate, 18
representative shear stress, 18
Reynolds number, 49, 117
rheogram, 17, 102
rheology, 1
rheopectic fluid, 7
roughness, 49, 50
Salmonella, 87
shear power intensity, 75, 79, 128, 132
  critical value, 77
shear rate, 3
  average, 22, 81, 104, 109
  in a pipeline, 8
  maximum in pumping, 12, *96*
  range for data collection, 10
shear rate correction factor, 11
shear stress, 3
  average, 18, 104
  in a pipeline, 8
shear work, 73, 74, 79, 128
  critical value, 77
  total, 80
shear-sensitive fluid, 72, 127
shear-sensitive particulates, 81
shear-thickening, 5
shear-thinning, 5, 97
slip, 19
stainless steel, 141
static volume, 76
steam properties, *145*
sterilizing value, 87
suction head, 48
system curve, 38, 113
system head, 38
thermal death time, 89
thermal processing, 55, 85
thixotropic fluid, 7
time-dependent fluid, 7
time-independent fluid, 6
torque curve method, 20
vapor pressure head, 48
velocity head, 38, 48
velocity head method, 36
velocity profiles, 55
viscosity, 1
  absolute, 4
  apparent, 5, 22, 97, 112
  kinematic, 4
volumetric average velocity, 50
water properties, 143
  steam, *145*
  viscosity, *147*
work, 38, 39, 41, 125
z value, 86